Praise for *The Rat*

'Utterly captivating . . . Be, ⸙ful humour, this is a thrilling adventure story of our own future'
Lewis Dartnell, author of *The Knowledge and Origins*

'Tom Chivers' meticulously researched book is intriguing, persuasive and eye-opening. His writing is warm and witty as he takes us on a surprisingly moving journey to decide – rationally – whether we really are playing Russian roulette with our future'
Hannah Fry, author of *Hello World*

'This book is about so much more than AI. It's about what happens when an attempt at perfectly rational behaviour meets the messy complications of humanity and its achievements. The brilliance of this book is the challenge it presents, because we can't examine the Rationalists without also examining ourselves. Tom Chivers is a fascinating and honest guide along that bumpy road'
Dr Helen Czerski, author of *Storm in a Teacup*

'Tom Chivers' book is like a self-help guide to stop panicking about technology for people who watched *Terminator* too many times when they were young. The content is completely gripping'
Ian Dunt, editor of politics.co.uk and author of *Brexit: What the Hell Happens Now?*

'A triumph . . . [Chivers] tries to assess ideas on a scale from true to false, rather than from quirky to offensive'
Scott Aaronson, author of *Quantum Computing Since Democritus*

'In this informative account of his encounters with the Rationalists . . . Tom Chivers follows a formula pioneered by those chronicling Silicon Valley for *Wired* magazine in the 1990s. He writes movingly about his insecurities over what the future will bring, telling the head of CFAR, "I'm scared for my children" . . . It is a resonant moment'
TLS

THE RATIONALIST'S GUIDE TO THE GALAXY

Tom Chivers is a science writer. He was *BuzzFeed* UK's science reporter between 2015 and 2018, and before that he spent seven years at the *Telegraph*, where he once interviewed Terry Pratchett and was told he was 'far too nice to be a journalist'. He has struggled on despite this handicap, being nominated for the British Journalism Award for science writing in 2017 and winning a Royal Statistical Society prize for statistical excellence in journalism in 2018, among other things. He tweets @TomChivers.

THE RATIONALIST'S GUIDE TO THE GALAXY

Superintelligent AI and the Geeks Who Are Trying
to Save Humanity's Future

TOM CHIVERS

WEIDENFELD & NICOLSON

First published in Great Britain in 2019 by Weidenfeld & Nicolson
This paperback edition published in 2020 by Weidenfeld & Nicolson
an imprint of The Orion Publishing Group Ltd
Carmelite House, 50 Victoria Embankment
London EC4Y 0DZ

An Hachette UK Company

1 3 5 7 9 10 8 6 4 2

A CIP catalogue record for this book is
available from the British Library.

ISBN (Mass Market Paperback) 978 1 4746 0879 4
ISBN (eBook) 978 1 4746 0880 0

Typeset by Input Data Services Ltd, Somerset

Printed and bound in Great Britain by Clays Ltd, Elcograf S.p.A.

www.orionbooks.co.uk
www.weidenfeldandnicolson.co.uk

For Billy and Ada. I hope the people in this book are right, and that you live to see humanity reach the stars.

Contents

PART THREE: THE WAYS OF BAYES

PART FOUR: BIASES

PART FIVE: RAISING THE SANITY WATERLINE

PART SIX: DECLINE AND DIASPORA

PART SEVEN: DARK SIDES

PART EIGHT: DOING GOOD BETTER

PART NINE: THE BASE RATE OF THE APOCALYPSE

Introduction

'I don't expect your children to die of old age'

Lord and Master! Hear me call. Oh, come the master!
Lord, the need is great! The ones I called, the spirits
Will not leave.

'Der Zauberlehrling', or 'The Sorcerer's Apprentice',
Johann Wolfgang von Goethe (1797). Translated using
artificial intelligence (specifically, Google Translate).

I was sitting in the passenger seat of a huge black BMW SUV, being driven around the byzantine freeways of the southern San Francisco Bay Area on a gorgeous October afternoon, when he said it: 'I don't expect your children to die of old age.'

To borrow a line from Douglas Adams,* when you're cruising down the road in the fast lane and you lazily sail past a few hard-driving cars and are feeling pretty pleased with yourself and then accidentally change down from fourth to first instead of third, thus making your engine leap out of your bonnet in a rather ugly mess, it tends to throw you off your stride in much the same way that this remark threw me off mine.

My companion was a guy called Paul Crowley, a man whose day job is as a cryptography engineer on Google's Android phone-operating system, but whose chief preoccupation in life is helping humanity reach the stars without first being destroyed by its own technological success.

---

* Specifically, from Chapter 13 of *The Hitch-Hiker's Guide to the Galaxy*, the bit where Ford introduces Arthur to Zaphod and Arthur says, 'We've met.'

There is a group of people, of whom Paul is one, who think that now is the crunch time. The next 100 years or so will be the inflexion point for humanity – either we go on and colonise the cosmos, becoming a galaxy-spanning civilisation of near-immortal demigods, or we annihilate ourselves with one or more of the technologies that we have developed. My children, they think, have a good chance of reaching demigodhood; they have, also, a good chance of not doing so. They want to improve the odds of the former.

I have been aware of these people for a few years. They're known as the Rationalists.

You've probably read a lot about artificial intelligence (AI) in recent years. Will it take our jobs? Will it form new, deadly, autonomous weapons on the battlefield? Will it lead to an era of inequality, as the rich buy all the robots and computers that run the new economy, and the poor find themselves left even further behind? Will we get those robot butlers we were promised?

These are serious and real concerns which deserve the many articles and books written about them, apart from the butler thing. But while the Rationalists are worried about that stuff, it's not the focus of their concern. Instead, they're worried that an AI will – in the relatively near future, the future that my children could easily live to see, or not far beyond it – become as smart as a human. And that when it does so, it will become as good as we are at designing artificial-intelligence systems, because designing artificial-intelligence systems is something that humans can do.

And so a machine that is as smart as a human could, possibly, very quickly improve itself, get better at improving itself, improve itself some more, and so on. An explosion would take place: suddenly, humans would find themselves vastly intellectually outgunned. Intelligence is what has made humans the most successful large animals on the planet; the tiny difference in DNA between us and gorillas, the thing that makes us smarter, is the difference that means our thronging billions live on every continent on the planet, while gorillas are going extinct in the mountains of Congo and Rwanda. If there's a machine that's smarter than us, the

Rationalists say, we would live only at its sufferance – as gorillas do, just about, at ours.

And, they say, just because a machine is smart, it doesn't mean that it's nice. It doesn't even mean that it's conscious. And if we aren't extraordinarily careful about how we build it – and even more careful about what we tell it to do – then it is possible that a future with AI could be, as far as we are concerned, extremely short and unpleasant. Or it could, equally, be glorious, spreading out across the galaxy, bounded only by the physical limits of entropy, light speed and the size of atoms.

This book is about that future. It's my attempt to work out whether I believe, as (some of) the Rationalists do, that we're on the brink of something – that my children, realistically, may not die of old age. It's also a look at the people themselves, who are fascinating, strange, clever, kind, frightened and self-sabotaging.

A word before I start. One of the people you'll meet in this book, the blogger and psychiatrist Scott Alexander, has an excellent habit. At the top of many of his blog posts – which are usually brilliant, thoughtful and terrifyingly long – he has a little line or two in italics, 'Epistemic status'; and then says how confident he is in his conclusions, and why. 'Epistemic status: Uncertain, especially on the accuracy of the economic studies cited.' 'Epistemic status: Pieced together from memory years after the event.' 'Epistemic status: Wild speculation.' I think this is wise.

So, here we go. *Epistemic status: Fairly confident.* I think most of the claims in this book are true; I think I have given a fair account to the best of my ability of what the people in it believe, and how they live. But I am human, and therefore have a brain that goes wrong in predictable ways. One of those ways (according to some more people you'll meet in these pages) is that when we find a fact we like, we ask ourselves, 'Can I believe this?', whereas if we find a fact we don't like, we ask ourselves, 'Do I have to believe this?'

I am fond of many – not all – of the people in this book, and I suspect that I therefore erred on the side of 'Do I have to?' when confronted with things that might make me like them less. For these reasons I am extremely confident that I have made errors throughout this book. I hope that none of them are major, or defamatory. Whether they are or not, they're mine, not anyone else's.

I flew to California in October 2017 to meet Paul and a few others involved in the Rationalist community. Their biggest in-real-life hub is based around Berkeley and Silicon Valley, although the community is really distributed around the internet.

This wasn't the first time I'd come into contact with them. I'd been aware of the community since about 2014, when I wrote a review of Nick Bostrom's *Superintelligence: Paths, Dangers, Strategies*. If you're vaguely aware of a conversation going on about whether or not AI will destroy the world, it's probably because of Bostrom's book. Elon Musk read it, in between making lots of money setting up PayPal and then systematically losing it again by trying to fly to Mars, and reported back: 'We need to be super-careful with AI. Potentially more dangerous than nukes.'[1] Bill Gates says we should all read it to understand AI.[2] Bostrom's work influenced Stephen Hawking's view that AI could be 'the best or worst thing to happen to humanity'.[3]

It was an amazingly dense, difficult book – writing my review, I opened it at random to select a passage, and ended up with: 'An oracle constructed with domesticity motivation might also have goal content that disvalues the excessive use of resources in producing its answers.'[4] I have a pretty good idea what that means, but it's not exactly a Ladybird Introduction to AI. Still, it sold extraordinarily well for what was essentially a work of academic philosophy, getting up to number 17 on the *New York Times* bestseller list.

And it is, once you get your head around it, somewhat terrifying. It compares humanity's efforts to build a superintelligent machine – and those efforts are ongoing, serious and, possibly, quite close to completion – to a bunch of sparrows trying to find an owl chick to raise, to protect them. As in, you can see where they're going with it, but it may be that they haven't 100 per cent thought through all the possible consequences.

Its release was, roughly speaking, when the Rationalists' concerns became mainstream. But apparently the book was not widely understood in the media, being met with a lot of references to *The Terminator*. A few people from the community, though, including Paul, read my review, and decided that I'd essentially got the gist of it. So they contacted me.

Over the next few years, I became more involved with the Rationalists. I started reading their websites; I learned the jargon, all these technical and semi-technical terms like 'updating' and 'paperclip maximiser' and 'Pascal's mugging' (I'll explain what all those things are later). I read the things you're supposed to read, especially and notably 'the Sequences' (I'll explain what they are later, as well). I came to terms with the huge possible impacts, positive and/or negative, of superhuman AI. And I became increasingly enamoured of their approach to the world, of which AI fears were only a part. It was also about people who want to make humanity better, to help us reach the stars, to stop us from destroying ourselves, to find ways of making us immortal. A whole related sub-branch is dedicated to making charitable giving more efficient. And it's about ways of helping us think about how we think – about using our best understanding of how the human mind works to make us better at achieving the things we want to achieve, and how to make us better at finding out things that are true and debating them with other people in charitable, kind ways.

I also gathered that human-level intelligence could be quite close: most people in the AI research field think it'll happen within the next century and possibly in the next few decades. And it all seemed pretty hard to argue with, on balance.

But somehow I hadn't put two and two together. The huge impacts and the possibly imminent arrival of AI were both things I understood and accepted on an intellectual level, but the implications of those two things hadn't really sunk in, in a visceral, gut-level, intuitive-understanding sort of way.

So when I was told that there was a real chance that my two young children would not die of old age, it shouldn't have shocked me – my children were, at the time, two and three years old; they could fairly confidently expect to live another 90 or 100 years; 90 years is well into the 'superhuman AI is more likely than not' bit of most researchers' predictions; superhuman AI, many people in the field believe, has the potential to either kill us all or make us near-immortal post-human demigods.

But it had all been an intellectual game, up to that point. Now we were talking about my actual, real-life children, my little toddlers Billy and Ada, who liked dinosaurs and *Octonauts* and the lower-quality Pixar movies. It left me somewhat winded. It brought to mind all the people who were worrying about robots taking people's jobs, or being used on the battlefield, and made me think: *The iceberg is 100 yards off the port bow, and you are worrying about whether the deckchairs are safe.*

As it happened, in the autumn of 2017 when I was out in California, there were a series of enormous wildfires.[5] A million acres of bush and forest in the north of the state burned; 43 people died and thousands of homes were destroyed, particularly around the Napa wine region. Those fires were just a few miles north of the Bay Area. Each morning when I woke up in my sad little Airbnb above a noisy nightclub in Berkeley I could smell woodsmoke; the sun was partly hidden behind a haze of it. When I went into San Francisco itself, doing the tourist thing, you could look north from Pier 39 and see that the far shore of the Bay was occluded behind a grey curtain of smoke, hanging in the valley like some ominous mist. One night I climbed the huge hill behind the Berkeley campus to see the sunset, and the smoke made the sun a vivid, bloody ball as it sank behind the almost invisible Golden Gate, 13 miles away: a startling sight which my iPhone camera was entirely unable to capture. (I even saw someone, up on the hill behind Berkeley, putting a cigarette out in the dry grass as he watched the sun sink into the smoke. I wanted to grab him by the collar and shout at him.)

A short distance away, everything was on fire and people were dying – but here, in this cosy little enclave of civilisation, no one was paying attention, even as the smoke drifted over their homes. The few who did were wearing surgical masks – in the face of a fiery death, people were worrying about asthma. Icebergs and deckchairs.

The metaphor is ridiculous, of course. There was no serious risk of anyone burning to death in Berkeley, but there *was* a pretty good chance of aggravating any pre-existing lung conditions. People were behaving perfectly sensibly. But I've been a journalist for over a decade now, and you don't get anywhere in this business by ignoring corny and obvious metaphors that are staring you in the face. So I started asking: Are we going to (metaphorically; possibly literally) burn to death?

It seemed to me important to find out whether the whole children-not-dying-of-old-age thing was a widely held belief. So I went and spoke to Anna Salamon.

Salamon is the president and co-founder of a non-profit organisation called the Center for Applied Rationality (CFAR, which people pronounce See-Far), and a key member of the Rationalists. CFAR, along with the Machine Intelligence Research Institute, or MIRI, is probably the closest thing the Rationalist community has to a real-world, as opposed to online, heart. Its office, which it shares with MIRI, is a couple of minutes' walk from the UCLA Berkeley campus, on a quiet road parallel with University Avenue, on the third floor of an unassuming office block. MIRI was set up by Eliezer Yudkowsky, the founder and driving force of the movement, an odd and polarising figure. (His name will come up again.) Salamon used to work for MIRI, on the problems of AI safety and existential risk, before going off to set up CFAR with the goal of training other bright, conscientious young people to work on the problems. Its mission is to instil in those bright young people the skills and methods of rationality that Yudkowsky and the Rationalists propound.

I'm a few minutes early to meet her and find myself looking around the shared lounge bit between the MIRI and CFAR offices, entirely alone and feeling extremely weirded out. It's a little dilapidated – it has something of a university junior common room feel, not the futuristic gleaming tech start-up look I'd been expecting. There's no reception, just a bunch of faintly elderly sofas and bean bags. One wall is dominated by a vast picture of the Earth from space; on another there is a whiteboard, covered in equations, as well as an H.P. Lovecraft-ish slogan ('Do not anger timeless beings with unspeakable names'), and a jaunty little reminder to 'Thank

Stanislav Petrov!' (Petrov, if you're unfamiliar with the name, was a Russian military officer who is credited with preventing a major nuclear war being triggered in September 1983, more of which later.)

There is also an expensive-looking road bike with drop handlebars and a pannier rack, propped up against a water cooler, with a post-it note saying: 'Is this your bike? Talk to Aaron.' For something with such a huge mission, this little place feels a bit lost in the vast suburban sprawl of the Bay Area; a little worn around the edges.

Before answering any question, Anna pauses for a tangible moment, a half-second or so; I'm fairly sure that this is a learned behaviour, an attempt to vet each statement to make sure it's something she *thinks*, rather than simply something she's *saying*. The conversation is initially quite hard. She's kind and thoughtful, but seems wary of me, and answers in short sentences or single words. I think she's concerned that I'm not here with good intentions as far as the Rationalist movement is concerned. (There is an understandable streak of paranoia among the Rationalists, I come to learn. Many of them are extremely intelligent and in some respects quite influential, but in others they are highly vulnerable – nerdy, often autistic or with other social deficits – and it would be extremely easy for me to write a book mocking them. I do not want to do that.)

I ask her, first, what her main goal is; CFAR is intended to train people with these rationality techniques, so I wondered whether that was an aim in itself, or whether it was a means to a greater end. She says that the primary goal is 'to help humanity reach the stars', and to do so while still recognisably human – not necessarily physically, but in terms of the things we care about and value. She thinks that there are lots of ways in which that might not happen, but the 'largest and most tractable part' of the problem is the risk of AI destroying humanity as we understand it.

Eventually I build up the courage to ask her the big one. Paul, I say, thinks that if humanity survives the next 100 years then we're probably going to make it to a glorious cosmic future. That my children probably won't die of old age.

'I agree with that.'
Either something terrible will happen, or they make it to . . .
'The singularity.'

Later on I meet Rob Bensinger, who's the research comms manager at MIRI. It is probably best to think of Rob as the aforementioned Eliezer Yudkowsky's messenger on Earth; Yudkowsky himself agreed only to answer technical questions, via email. Rob, a polymath who had become part of Yudkowsky's circle a couple of years before, speaks for him, like one of the angels who appear in the Bible when God has something to say but can't bring Himself to turn up in person.

When I meet Rob, MIRI is preparing for a 'retreat'. They weren't actually going anywhere, but in recent years they'd found that when they had previously had real, proper, go-off-into-a-cabin-in-the-woods-somewhere retreats they had been extremely successful. They'd really focused people's minds, improved their productivity.

But hiring a cabin that can fit a dozen people is quite expensive and inconvenient, so they were experimenting with ways of having the same effect without needing to go anywhere. Instead, they'd hung a large white sheet across the office and used uplighting and other little visual tricks to make it feel like somewhere else. Rob said that it had been effective. It struck me as a rather clever solution.

I ask him the children question, and he demurs. 'I wouldn't like to be on the record as saying something that specific. It's a pretty sensitive thing and I wouldn't want to off-the-cuff it.' I press him a bit, though, and he says that Paul's view 'seems normal to me'. 'Most people expect AI this century, and most people interested in AI risk generally think the risk is pretty serious, not just a small risk but a medium-sized to large risk, shall we say. I'd say it's a this-century problem, not a next-century problem. I think most people will agree with me, within AI. The Open Philanthropy Project (OpenPhil) gives it at least a 10 per cent chance of happening in the next 20 years.'

As it happened, I was going to OpenPhil the next day. It's across the Bay from Berkeley, in downtown San Francisco itself.

OpenPhil and GiveWell are two organisations run on Rationalist lines that look at the most effective ways to donate money to charity. They're central to something called the Effective Altruism (EA) movement, which is strongly linked to the Rationalist community. OpenPhil in particular has donated millions of dollars to AI safety organisations, and MIRI in particular, over the years.

Holden Karnofsky, the co-founder of both OpenPhil and Give-Well, confirms what Rob said, that OpenPhil thinks there's about a 10 per cent chance of 'transformative' AI in the next 20 years. 'That would clearly meet your criteria' of my children not dying of old age, he said. Isn't that terrifying? I ask him. 'Yes,' he says. 'We live in a truly weird time.' We don't realise how fast things are changing now, in a way unlike any other time in history, but it's only going to get faster: 'If transformative AI comes, that could be transformative in ways that would make the Industrial Revolution look small. Yes, it's really strange and it's disorienting.'

Not everyone I spoke to agreed with this. Some people reckoned the timeline was too short, and that it was unlikely (though not impossible, by any means) that human-level AI would arrive in my children's lifetimes. Others thought that the timeline was perfectly realistic but that human-level AI wouldn't bring the sort of spectacular change (and possible destruction) that would lead to them not dying of old age. Others were understandably wary about putting these sorts of numbers on things to begin with.

But it seemed like we were at least dealing with something that was *not unrealistic*. Sensible, intelligent people, including AI researchers at serious AI companies, senior academics and so on, thought that there was a respectable chance that the next 100 years would see either an ascension to demigodhood or a complete, civilisation-ending catastrophe. More than civilisation-ending, in fact: human-life-ending.

This book is about some of those people. Specifically, it is about a community of them who came together around a series of blog posts written by Yudkowsky in the mid- to late 2000s, and who are known as the Rationalists.

It is also, in part, an attempt to work out whether I agree with them.

Part One

Introductions

Chapter 1

Introducing the Rationalists

The Rationalist community, as it exists now, is sprawling and global. It has hubs in a dozen or more cities and a thronging online presence. It's full of strange people with strange ideas – about AI (the idea that AI has the potential to be an existential risk to humanity can, I think, be largely traced back to it, or its precursors), about transhumanism, cryonics; about the universe being a simulation – and unorthodox practices, such as polyamorous relationships (ones with several people at once) and group living, which have led to outsiders accusing it of being a cult.

This whole ecosystem has its roots in the writing of the strange, irascible and brilliant Eliezer Yudkowsky. The key text – the holy book, according to those who think the whole thing is a quasi-religion – is a huge series of blog posts he wrote in the mid-2000s, an ambitious, sprawling set of writing which takes in everything from evolutionary biology to quantum mechanics to AI, and which came to be known as the Sequences. But as far as I can tell, the first visible sign of its birth is a single, much older blog post written on 18 November 1996. It was entitled 'Staring into the Singularity'.[1] Yudkowsky was 17 years and two months old at the time. The post is still online, by the way. But it is marked at the top by a big red warning triangle, like the sort you get in the back of your car to warn of a road accident, saying: 'This document has been marked as wrong, obsolete, deprecated by an improved version, or just plain old.' This sort of thing goes on a lot in the Rationalist community. Being wrong is actively praised, as long as you hold up your hands, admit it and correct it.

'Staring into the Singularity' is a fascinating read, although the logic doesn't really bear scrutiny, as you'd expect from a

17-year-old. It begins: 'If computing speeds double every two years, what happens when computer-based AIs are doing the research?' This is a reference to Moore's law, which says (roughly, in one formulation) that computers get twice as powerful every two years. So computing speed doubles every two years. Yudkowsky points out a corollary: that two years for a human need not seem like two years for a computer. 'Computing speed doubles every two years of work. Computing speed doubles every two *subjective* years of work.' That is, if a computer can think as powerfully as a human, but twice as fast, then it can do two years' worth of work in a single year. 'Two years after Artificial Intelligences reach human equivalence, their speed doubles. One year later, their speed doubles again. Six months – three months – 1.5 months ... Singularity.' Things would speed up, exponentially. The world would be changing too fast for us to understand the changes. We would be through the looking glass. This is, roughly, the idea of the 'fast take-off' that Nick Bostrom would describe nearly two decades later, although Yudkowsky's version makes a few weird assumptions and leaps of logic (as, again, is fair enough given his age).

The term 'singularity', by the way, is a reference to physics and black holes. When an object is massive enough and small enough, it bends spacetime so much that the usual laws of physics no longer work. By analogy, when intelligent systems start improving themselves fast enough, our usual ways of predicting the future – our assumptions that tomorrow will be essentially like today – will, say singularitarians, break down. The computer scientist and science-fiction writer Vernor Vinge wrote in 1983: 'We will soon create intelligences greater than our own. When this happens, human history will have reached a kind of singularity, an intellectual transition as impenetrable as the knotted space-time at the centre of a black hole, and the world will pass far beyond our understanding.'[2]

Yudkowsky's explicit goal in 'Staring into the Singularity' is to bring about AI – the singularity – as soon as possible. 'Human civilisation will continue to change until we either create super-intelligence, or wipe ourselves out,' he wrote. This superintelligent thing we create, he thinks, can solve all of humanity's problems, and it is high time that it does so: 'I have had it. I have had it with crack houses, dictatorships, torture chambers, disease, old age, spinal paralysis, and world hunger. I have had it with a planetary death rate of 150,000 sentient beings per day. I have had it with this planet. I have had it with mortality. None of this is necessary. The time has come to stop turning away from the mugging on the corner, the beggar on the street. It is no longer necessary to look nervously away, repeating the mantra: "I can't solve all the problems of the world." We can. We can end this.' Yudkowsky was, you suspect, quite an annoying 17-year-old, but he was undeniably bright.

In 2000 – still only 20, remember – he founded the Singularity Institute for Artificial Intelligence. The Singularity Institute was a little non-profit based in Berkeley which would later become known as the Machine Intelligence Research Institute, which you've already met. It had, at first, the goal of bringing about this glorious technological future, and Yudkowsky had set a target date for achieving the singularity. It was 2005. (He didn't manage it.)

I asked Paul Crowley – whom you met in the introduction, driving me around northern California – about all this. 'My broad picture of how it started,' says Paul, 'is Eliezer started by thinking superintelligence is the key to everything, and we need to get there as quickly as possible. *It's intelligent*, he thought, *so it'll do the right thing.*'

But even by the time he founded the Singularity Institute, at least according to what he wrote later,[3] Yudkowsky had started to wonder whether he was making a terrible mistake. 'The first crack in my childhood technophilia appeared in, I think, 1997 or 1998,' he wrote, when he noticed his fellow techno-enthusiasts being glibly optimistic about the difficulties of controlling future technologies – specifically, nanotech. By the end of that debate, he says, the young Yudkowsky 'had managed to notice, for the first time, that the survival of Earth-originating intelligent life stood at risk'. Still, he cracked on with the Singularity Institute, full steam ahead, for superintelligence. 'Just like I'd been *originally* planning to do,' he writes, with some scorn, 'but now, with a *different reason*.'

This feels, reading his work now, like a key moment, at least in retrospect. What Yudkowsky didn't do, he says, was 'declare a Halt, Melt and Catch Fire'. He didn't look at his own thinking, accept that it was all completely wrong and cast out his conclusions (in his own, later, words, he didn't appreciate 'the importance of saying "Oops"'). Instead, he looked at his thinking, realised that it was wrong, and decided that his conclusions were conveniently right anyway.

But slowly – between 2000 and 2002, between the ages of 20 and 22, probably too young to be placing the entire future of the world on your own shoulders, but that appears to have been the sort of person young Yudkowsky was – he came to the realisation that not only was he wrong, he was disastrously wrong. The exact ways in which he was wrong are going to be the topic of much more discussion in this book, but according to Yudkowsky himself by the time he had reached the wise old age of 27 his stupidity had led him to try to build a device which would destroy the world. '[To] say, *I almost destroyed the world!*, would have been too prideful,' he wrote.[4] But he had been *trying* to do something which he thought, if he had had the wherewithal to actually do it, would have done exactly that.

So he decided to try to save the world instead.

Yudkowsky was not the first person to think about what would come after humans. He was firmly part of the traditions of trans-humanist and singularitarian thinking, which had been around for years when he was writing 'Staring into the Singularity'; some of the ideas they hurled about had existed for millennia. Bostrom notes in a paper that in the *Epic of Gilgamesh*, a 4,000-year-old Sumerian legend which foreshadows parts of the Old Testament, 'a king sets out on a quest for immortality. Gilgamesh learns that there exists a natural means – an herb that grows at the bottom of the sea.'[5] (He finds it, but a snake steals it off him before he can eat it, as is so often the way.) The Elixir of Life, the Philosopher's Stone, the Fountain of Youth and various other myths represent similar ideas.

Bostrom also points out that early transhumanist-style myths contain an element that remains in modern discussion of its ideas: hubris leading to nemesis. Prometheus steals fire from the gods, which most of us can agree was a good thing from humanity's point of view: his punishment was to have his liver repeatedly pecked out by an eagle for eternity. Daedalus improves on human abilities by, among other things, building wax-and-feather wings to grant himself and his son Icarus the power of flight; Icarus promptly flies too close to the sun, melting the wax, and plunges into the sea. St Augustine thought that alchemy, and the search for a panacea or eternal life, was ungodly, possibly demonic.

The idea that science could improve on the human-basic form became more plausible after the Enlightenment and Renaissance. Nicolas de Condorcet wondered in 1795 whether science would progress until 'the duration of the average interval between birth and wearing out has itself no specific limit whatsoever', and that people would choose to live until 'naturally, without illness or accident, [they find] life a burden'.[6] Benjamin Franklin wrote of wanting to be 'embalmed' in such a way that he could be revived in the future, since he had 'a very ardent desire to see and observe the state of America a hundred years hence'.[7] Bostrom points out that this foreshadows the modern idea of cryonics, preserving the brain for revival in the future.

The term 'transhumanism' and some of its most recognisable ideas sprang up in the first half of the twentieth century. In 1923 J.B.S. Haldane predicted a world in which humans used genetic science to make themselves cleverer, healthier and taller. The term itself was apparently coined by Julian Huxley, brother of Aldous, in 1927: 'The human species can, if it wishes, transcend itself – not just sporadically, an individual here in one way, an individual there in another way – but in its entirety, as humanity. We need a name for this new belief. Perhaps *transhumanism* will serve'.[8]

But transhumanism and singularitarianism really took off as philosophies in the last decades of the twentieth century. There were various different, and to some degree competing, ideas of what transhumanism involved – Yudkowsky, in a since-deleted online autobiography[9] he wrote at the age of 20, credits Ed Regis' 1990 book *Great Mambo Chicken and the Transhuman Condition*, an early, comical taxonomy of these different visions, as an inspiration. The idea of cryonics began to become more popular in this period – super-cooling the brains (and perhaps bodies) of dying people in order to preserve them, as Franklin wished, with the idea of reviving them when technology advanced sufficiently to do so. Transhumanists also talked about how nanotechnology can transform everything. A large subset of them were keen on the idea of uploading – scanning a human brain so precisely, probably by slicing it apart, that you could simulate it in a computer, creating a digital version of the mind that you scanned. (The original, of course, would be destroyed in the process.) Machine-brain interfaces – ways of linking a human brain to a computer, or linking human brains via computers, to improve human cognition – were a constant topic. All of this, naturally, overlapped with the 'singularitarian' vision of a world in which superintelligent AI or other technological advances rendered human life unrecognisable (but unrecognisable, they'd have said, in a *good* way). Most of all, they wanted – want – to stop death. About 150,000 people die every day, worldwide. Most of us wave that away, saying that death gives life meaning, or that eternity would be boring. The transhumanists (not unreasonably, to my mind) ask: OK, but if death didn't exist, would you all be saying, 'We ought to limit our lives to about 80 years, to give them meaning?'

As befits a movement that gets a book written about it with the term 'Great Mambo Chicken' in its title, some of its members were and are – by the tightly corseted standards of Western society, I should say – deeply weird. There's an affectionate 2006 *Slate* article about transhumanists which says, at one point, 'Remember those kids who played *Dungeons & Dragons* and ran the science-fiction club in your high school? They've become transhumanists.'[10] There appears to have been an element of truth in that gently mocking phrase.

Transhumanists have a tendency, for instance, to give themselves strange names. The *Slate* article mentions one who calls herself Wrye Sententia (Dr Sententia is a professor at UC Davis and the director of a non-profit called the Institute for Ethics and Emerging Technologies. Having a strange name doesn't stop you doing interesting work.) Another changed his name from Fereidoun M. Esfandiary, which was an interesting enough name to begin with, to FM-2030. There's a Tom Morrow, which is lovely. And there's a guy who was once called Max O'Connor but who changed his name to Max More, because 'It seemed to really encapsulate the essence of what my goal is: always to improve, never to be static. I was going to get better at everything, become smarter, fitter, and healthier.'[11]

More would later become CEO and president of Alcor, one of the largest cryonics companies in the world. But he is mainly relevant to this story because in 1988, along with Tom Morrow (see? Lovely), he began publishing *Extropy Magazine*. It was mainly about transhumanism – how to improve upon the human form, make it immortal, make it cybernetic, and so on. In 1992 he founded something called the Extropy Institute, which set up a mailing list – a sort of early precursor of social media, for those of you under the age of 35; you just all chat in your emails – called the Extropians.

One of the names on the Extropians' mailing list was Eliezer Yudkowsky. 'This was in the 1990s,' says Robin Hanson, an economist at George Mason University and an important early Rationalist figure. 'Myself, Nick Bostrom, Eliezer and many others were on it, discussing big future topics back then.' But neither Bostrom nor Yudkowsky were satisfied with the Extropians. 'It was a relatively libertarian take on futurism,' says Hanson. 'Some people, including Nick Bostrom, didn't like that libertarian take, so they created the World Transhumanist Association, explicitly to no longer be so libertarian.' The World Transhumanist Association later became Humanity+ or H+. 'It hardly trips off the tongue as a descriptor,' says Hanson. 'But that's what they insisted they call everything.' Humanity+ had a more left-wing, less utopian approach to the future.

Yudkowsky, on the other hand, felt that the problem with the Extropians was a lack of ambition. He set up an alternative, the SL4 mailing list. SL4 stands for (Future) Shock Level 4; it's a reference to the 1970 Alvin Toffler book *Future Shock*.[12] Future shock is the psychological impact of technological change; Toffler describes it as a sensation of 'too much change in a short period of time'.

Yudkowsky took the concept further, dividing it up into 'levels' of future shock, or rather into people who are comfortable with different levels of it. Someone of 'shock level 0' (SL0) is comfortable with the bog-standard tech they see around them. 'The use of this measure is that it's hard to introduce anyone to an idea more than one shock level above,' he said. 'If somebody is still worried about virtual reality (low end of SL1), you can safely try explaining medical immortality (low-end SL2), but not nanotechnology (SL3) or uploading (high SL3). They might believe you, but they will be frightened – shocked.'

He acknowledged that transhumanists like the Extropians were SL3, comfortable with the idea of human-level AI and major bodily changes up to and including uploading human brains onto computers. But he wanted to create people of SL4, the highest level. SL4, he says, is being comfortable with the idea that technology, at some point, will render human life unrecognisable: 'the total evaporation of "life as we know it"'. (I'm taking this from a 1999 post[13] of his on SL4, when he'd just turned 20. He also fleshes it out in a long essay called 'The Plan to Singularity'[14] from about the same time.) He wanted to convert SL2s and SL3s to SL4s, to build a community of people who were comfortable talking about ideas of the post-human future. So he set up this mailing list and called it, reasonably enough, SL4.

Its archives are still available online, and digging through them is a fascinating experience. It's a bit like that Sex Pistols gig in 1976, where there were only 40 people in the audience but all of them went on to form major bands. Going through the list of authors, you find the founders of major AI companies – such as Ben Goertzel – or AI researchers like Bill Hubbard. Wei Dai, an AI researcher at Imperial College London who played an important role in the creation of cryptocurrencies, is on there. Bostrom and Hanson are both there, and Anna Salamon. Other people who play roles in the story – Michael Vassar, Michael Anissimov – are contributors.

Nick Bostrom did a minor double-take when I asked him about SL4 and the Extropians, as though he hadn't thought about it in a long time. I think he gave a sort of chuckle. 'Yeah, it was humble beginnings,' he said. 'I'd been thinking through some of these things before, but I didn't know there were other people thinking about it. It's a bit strange. Nowadays you'd just Google it and immediately find whatever there is, but in the early 1990s when I was a student no one else was interested in it. So it was a bit of a revelation when I started using the internet in 1996 that there were these communities, people chatting about it.'

Several of the key concepts that do the rounds in the Rationalsphere these days first arose on SL4 and the Extropians. The aforementioned 'paperclip maximiser' was first mentioned there, possibly by Yudkowsky: 'Someone searched [the Extropians' archive] recently and found a plausible first mention by me,' he told me by email; he was and remains wary about talking to me on the phone. 'I wasn't sure if it was me, Nick, or Anders Sandberg, but it kind of sounds like me.' The 'AI box' experiment, in which Yudkowsky attempted to demonstrate that even an 'oracle' super-intelligent AI, locked in a box and only able to communicate by text, was not safe, took place on SL4.[15] Bostrom first linked to his paper arguing that we may be living in a computer simulation on SL4.[16]

But although SL4 gathered quite an impressive bunch of people, it still wasn't enough to satisfy Yudkowsky. Looking at the archives, you see that he's extremely busy in the first few years, up to about 2004, but later on he seems to be less involved. No new threads of his appear at all between 2005 and 2008. The Rationalists' own semi-official history of themselves, on the wiki page of Yudkowsky's website LessWrong, says that he 'frequently expressed annoyance, frustration, and disappointment in his interlocutors' inability to think in ways he considered obviously rational' and that after 'failed attempts at teaching people to use Bayes' Theorem, he went largely quiet from SL4 to work on AI safety research directly'.[17]

Then Robin Hanson, the economist and fellow SL4/Extropians commenter, set up a blog of his own called Overcoming Bias. 'I started this blog after I got tenure at George Mason,' Hanson told me, 'as something to do in my spare time.' When I speak to him, via Skype from his office at the university, he'd accidentally left his window open all weekend during one of the more dramatic periods of cold weather the US East Coast had seen for a while. He was wrapped up in a puffa jacket and woollen hat and his breath was visible in the air, even on the low-res Skype connection. 'I decided to theme it on overcoming bias.'

This was 2006, a few years before the publication of Daniel Kahneman's famous book *Thinking, Fast and Slow*, about the various systematic biases in human thought. But Kahneman's groundbreaking work with Amos Tversky was already extant and slowly becoming more widely known. Hanson, a polymath and autodidact in a similar, if less extreme, vein to Yudkowsky, had picked up a lot of Kahneman and Tversky's work in his travels around the sciences – he'd qualified as a physicist, before doing post-grad degrees in social science and economics. Overcoming Bias was explicitly founded on the 'general theme of how to move our beliefs closer to reality, in the face of our natural biases such as overconfidence and wishful thinking, and our bias to believe we have corrected for such biases, when we have done no such thing'.[18]

He invited a few old Extropians/SL4 veterans to come and join him, people who'd impressed him with the quality of their thinking. Among them were Nick Bostrom and Eliezer Yudkowsky. 'Nick just blogged a few things,' he said. 'But Eliezer blogged a lot, which was great.' It's at this point that Yudkowsky began what would later become known as the Sequences. In essence, they were a reaction to the fact that he couldn't get people to understand what he was talking about when he said that AI was a threat.

The problem he had was that no one really took him seriously. So in order to explain AI, he found he had to explain thought itself, and why human thought wasn't particularly representative of good thought. So he found he had to explain human thought – all its biases and systematic errors, all its self-delusions and predictable mistakes; he'd found a natural home on Overcoming Bias. And to explain human thought, he found he had to explain – everything, really. It was like when you pull on a loose thread and end up unravelling your entire favourite jumper.

It was a meandering, unfocused thing, for a long time; at one point he gets on to quantum physics; at another he approvingly cites George Orwell's (somewhat silly) proscriptions against using the passive voice. Paul Crowley tells an illustrative story. 'There's this post, about fake utility functions,'[19] he says. (Don't worry about what a utility function is or how it can be fake.) 'If you want to know the story of how this got written, it's a good one to read. It begins by saying something like, "Today I can finally talk about this idea of fake utility functions. I was going to talk about it six months ago. But then, when I sat down to write it, I found I had to set out this idea, and then to explain that idea it helps if I explain this other idea. And then I thought it would be easier if the reader understood evolutionary biology, so I ended up writing an introduction to evolutionary biology." He ended up writing about two dozen posts just on evolutionary biology. And the joke of it is this post wasn't even some cornerstone of the whole thesis; it was just something he wanted to write.'

Slowly, the blog posts built up, and up and up. For an idea of how much, think of *The Lord of the Rings* books. When you add all three together, they come to about 455,000 words. *War and Peace*, a book which is actually more famous for being long than it is for being good (it's OK), is about the 587,000 mark. According to the Kindle app on my iPhone, *War and Peace* is 18 arbitrary dots long. *Rationality: From AI to Zombies*, the edited e-book edition of Yudkowsky's blog posts, merits 19 dots. If the Kindle app's length indicator is accurate, then that puts *RATZ* at around 620,000 words long.* The unedited Sequences were more like a million. That's a fair old slog. There are few things so dispiriting as reading on a Kindle and realising that after 30 minutes you've only gone from 3 per cent to 4 per cent. And this is not a book about elves fighting orcs, which if nothing else keeps you moving along. Yudkowsky is an engaging writer, but by its nature it's heavy going.

---

* The Bible still wins. The King James Authorised Version weighs in at 783,137 words. While looking that up, I learned that in the early days someone managed to miss one word out, getting it down to a more manageable 783,136 but unfortunately changing the Sixth Commandment to read 'Thou shalt commit adultery'.

But it became pretty successful. In 2009 Yudkowsky moved his blog posts over to a new website, LessWrong, which was intended as a sort of community hub where anyone could post. At about the same time – 2010 – he started publishing something else, his Harry Potter fan fiction *Harry Potter and the Methods of Rationality*, which does exactly what it says on the tin: it involves a nerdy scientist-Harry trying to work out what the rules of this magical universe are, using Rationalist-style methods. It was a surprising success, gathering 34,000 reviews on the site FanFiction.net; it may be the most-read thing that Yudkowsky has ever done, and attracted large numbers of readers to his other work, especially LessWrong.

At its peak, LessWrong had about a million page views a day.[20] Some posts had hundreds of thousands of unique page views (a metric that avoids the problem you get of someone clicking 'refresh' and suddenly counting as two hits). It's probably not completely inaccurate to say that a million people have read *some* of the Sequences, and I'd guess that the number of people who've read the whole thing is probably a high-five figure or low-six. I may be off by an order of magnitude, of course – there's no easy way to tell.

What Yudkowsky was trying to do with all this was to explain why AI was dangerous. But because he found that first he had to describe intelligence, and human intelligence, his project became more ambitious: to improve human rationality, in order to help prevent humanity from destroying itself.

Chapter 2

The cosmic endowment

We'll get on to why the Rationalists think that AI is so dangerous soon. But first we should look at why they, and the singularitarians who came before them, are also so keen on it. The gamble, they think, is between extinction and godhood.

According to the Rationalists, getting AI right could be the greatest thing that ever happens to our species. If humanity survives the next few decades, or maybe centuries – it's not clear exactly how long, but probably a fairly insignificant period in comparison to how long we've already existed, and *certainly* an insignificant period in comparison to how long everything else has – then things could go extraordinarily right for us. This is what Paul Crowley meant, or part of what he meant, by saying that he didn't expect my children to die of old age.

It is improving technology, and specifically AI, that people are talking about when they refer to this glorious future. 'The potential benefits are huge, since everything that civilisation has to offer is a product of human intelligence,' wrote the authors of an open letter in 2015.[1] '[We] cannot predict what we might achieve when this intelligence is magnified by the tools AI may provide, but the eradication of disease and poverty are not unfathomable.' That letter was signed by more than 150 people, including dozens of senior computer scientists and AI researchers, three founding members of Apple, Google DeepMind and Tesla, and Professor Stuart Russell of Berkeley, the author of the standard textbook for AI undergrads. (The late Stephen Hawking also signed, but AI researchers used to get understandably annoyed when he made the headlines rather than the people who actually do this for a living.) Max Tegmark, a professor of cosmology at the Massachusetts

Institute of Technology and director of its Future of Life Institute, writes in his book *Life 3.0: Being Human in the Age of Artificial Intelligence* of 'a global utopia free of disease, poverty and crime' as a possible outcome of the development of a powerful AI. These are real, serious people who believe that, in the reasonably foreseeable future, AI could solve some of humanity's most pressing problems. But 'solving our problems' is actually the least of it. If humanity survives, we have to start looking at some very big numbers.

Let's imagine the Earth will probably be able to support human life for another billion years or so. (At around that point, the sun will enter a phase in which it is much brighter and hotter than it currently is; it will cause the Earth to enter a runaway greenhouse process as the seas evaporate, and it will become too hot for complex life.[2]) Let's imagine that humans continue to live for a century or so each for the next billion years, and that the human population settles at a nice, sustainable 1 billion, less than one-seventh of its current levels. These are the assumptions that Nick Bostrom – author of the aforementioned *Superintelligence*, and founder of Oxford's Future of Humanity Institute (FHI) – goes with.[3] That would mean that we would have at least 10,000,000,000,000,000 descendants. The total number of *Homo sapiens* who have ever lived up to now, according to an estimate by the Population Research Bureau, is about 108,000,000,000.[4] In other words, the entire history of humanity so far represents only about one-ninety-thousandth of what it could be, if we just avoid being wiped out.

But! We *still* have only just started to scratch the surface. What if humanity leaves the Earth? Imagine a 'technologically mature' civilisation, says Bostrom. One that can build spacecraft that travel at 50 per cent of the speed of light. That civilisation could reach 6,000,000,000,000,000,000 stellar systems, he calculates, before the expansion of the universe puts the rest out of reach. One that could travel at 99 per cent of the speed of light could reach about 15 times as many as that.[5] Imagine that 10 per cent of those suns have planets that are or could be made habitable, and on average could each sustain 1 billion people for 1 billion years. That would put the number of humans who could exist in the future at around

10^{35}, or 1 followed by 35 zeroes. All the humans who have ever lived would be vastly less than a rounding error, compared to the ones who could follow us, if we get it right.

But! Yes. It gets bigger. Much bigger.

First, we could build our own habitats (an example would be the Orbitals in Iain M. Banks' Culture novels, thin wheels of matter millions of kilometres in diameter, with humans living on the inside of the rim), out of spare space rocks, so we're not limited by the number of planets we happen to find. That gets Bostrom up to 10^{43} potential humans.

And *then* we could think about what happens when we start uploading human minds into computers. Then we are far less limited by space. Humans would need, instead of an appreciable fraction of the surface of a planet, a few square picometers of circuitry. Bostrom throws some plausible-sounding numbers in there about how dense you can make your hardware, how much energy you can get from a given star, and how many computations per second are required to simulate a human mind, and comes up with a *lower bound* – a conservative, worst-case estimate – of 10^{58} possible human lives of 100 years each.

'One followed by 58 zeroes' may sound like a meaningless Big Number to you, and it does to me, but it is *extraordinarily* vast. 'If we represent all the happiness experienced during one entire such life with a single teardrop of joy,' says Bostrom, 'then the happiness of these souls could fill and refill the Earth's oceans every second, and keep doing so for a hundred billion billion millennia.'[6] Does that give a more visceral sense of how enormous it is? I don't know if it does. If not, just remember that even compared to the ridiculously vast numbers that astronomers throw around from time to time, this is seriously huge.

Bostrom might have his numbers wrong, of course. He has done his best to think conservatively, but when putting the numbers together like this he could easily be off by orders of magnitude. But even if you knock off six orders of magnitude and say there's only a 1 per cent chance of it being correct anyway, then even 'reducing existential risk by a mere one-billionth of one-billionth of one percentage point'[7] can be expected to do as much good, in terms of years of life saved, as stopping 100 quintillion actual people from dying. That's one followed by 20 zeroes.

So, yes, Bostrom's maths could be badly wrong. He could have got his figures off by a factor of 10,000, or a million, or 10 billion. And yet, *all the good done right now by every charity in the world would still be a drop in an ocean that is itself a drop in a much bigger ocean* in comparison to the good that would be done by slightly reducing the chance that humanity gets destroyed before it can take to the stars.

You might want to reject these numbers out of hand because they're weird and they give you weird results. That's not actually a stupid thing to do, according to the Rationalists: there is a thing Yudkowsky came up with, called Pascal's Mugging,[8] related to the famous wager which says that if you simply multiply risk by reward, you're vulnerable to absurd situations. In Bostrom's rather whimsical version of it,[9] the example is that a mugger comes along and demands Pascal's wallet. Pascal points out: 'You have no weapon.' 'Oh good point,' says the mugger. 'But how about if you give me the wallet, I come back tomorrow and give you 10 times the value of the money in it?' 'Well', says Pascal, 'that's not a very good bet, is it. It's hugely likely that you'll just not come back.'

But the mugger then says: 'Actually, I'm a wizard from the seventh dimension. I can give you any amount of money you like. I can give you, in fact, any amount of happiness you like. Let's say that the money in your wallet could buy you one happy day. [Assume for the sake of argument that money can buy happiness.] And let's say that you think there's only a 1 in 10^{100} chance that I'm telling the truth. Well, in that case, I'll offer you $10^{1,000}$ happy days.'

By a utilitarian calculus – the idea that you should multiply the chance of something happening by the reward it would bring if it does, exactly the sort of reasoning that Bostrom uses to think about the cosmic endowment, or for that matter that investors and gamblers use to determine where to put their money – this is a good bet. If Pascal took it, on average, he'd expect a 10^{990}-fold return on his investment. But it is, also, pretty obviously ridiculous.

The wizard-mugger can just keep upping the numbers he offers until it *becomes* a good bet.

So it's OK to be wary; you should be, when someone comes up and mouths a lot of maths and numbers and technical talk that you can't follow but which they say supports their point. The Rationalists have a term for that, in fact: 'getting Eulered',[10] blinded by numbers. But that doesn't mean you should simply dismiss it. If you can't follow the maths, you should be wary, but you should *try* to follow the maths. One of the founding principles of the Rationalist movement is that, as Scott Alexander puts it, 'when math tells you something weird, you at least consider trusting the math. If you're allowed to just add on as many zeroes as it takes to justify your original intuition, you miss out on the entire movement.'[11] A weird-seeming answer is a warning flag, rather than a stop sign: a thing to investigate rather than reject.

And, having investigated for a decade or more, the Rationalists are pretty confident in their numbers – which is why they, and the Effective Altruism movement which is closely aligned with them, are concerned about AI risk. They see the reward of surviving the next century or so as potentially enormous, and AI as one of the – if not the – most likely things that will stop us doing that.

Part Two

The Paperclip Apocalypse

Chapter 3

Introducing AI

In this chapter, we're going to get together a working definition of what AI actually is, before we discuss the reasons why the Rationalists think AI could go so wrong.

There are lots of things around right now that are described as 'AI'. But they're all what is known as *narrow* AI. For instance, chess-playing AIs are extremely good at chess, but clueless about everything else. They can't help you with your tax return or remember to feed your cat. Google Maps is pretty good at working out optimal routes from A to B, and only rarely directs you through the North Sea or whatever, but it doesn't know a queen's gambit from the Sicilian defence. Humans, on the other hand, can apply themselves to learning ballroom dancing, or the guitar, or chemistry, or poetry-writing – apparently some other things as well. What the Rationalists are concerned about, broadly speaking, is the development of artificial *general* intelligence, or AGI: a computer that can do all the mental tasks that we can.

At this point I really, really want you to put all pictures of *The Terminator* out of your heads. That's important. It won't help. Nothing we discuss will be made clearer by images of grinning metal robots or Skynet achieving self-awareness in August 1997. I'm *fairly* sure this trope irritates most Rationalists, but I'm damn sure that I'm tired of it myself after a year of conversations in which someone asks me what the book is about, I say 'artificial intelligence destroying the world', and they nod sagely and say, 'Ah, Skynet.' So. Please. *The Terminator*: forget it.

But you might wonder what I *do* mean when I talk about 'artificial intelligence'. You might, for instance, reasonably point out that there is a fairly large disagreement about what 'intelligence' means, even before you start talking about whether it's artificial or not. Conveniently for me, the standard textbook of AI, *Artificial Intelligence: A Modern Approach* (*AIAMA*), by Stuart Russell and Peter Norvig, tries to answer this question. It divides up possible approaches and definitions by whether they talk about *reasoning* or about *behaviour*, and whether they attempt to reason/behave like a *human* or whether they attempt to reason/behave *rationally*.[1]

A computer that behaves humanly is the sort of thing imagined in the old Turing test. In 1950 the British scientist and code-breaker Alan Turing, apparently bored with debate over whether 'machines can think', wrote in a paper in *Mind* that 'If the meaning of the words "machine" and "think" is to be found by examining how they are commonly used, it is difficult to escape the conclusion that the meaning of and answer to the question "Can machines think?" is to be sought in a statistical survey such as a Gallup poll.'[2] Instead he proposed a simpler, more unambiguous test, the 'imitation game'. A human interviewer holds conversations with two interlocutors, whom he cannot see. One is a human; one is an AI. The interviewer can ask whatever questions he likes; if he cannot reliably tell one from the other, said Turing, then to all intents and purposes we should treat it as a thinking thing.

Turing's famous test is rightly held up as a great pioneering work, and has the enormous advantage that it sidesteps grimly philosophical questions such as 'Is it conscious?' and puts a simple, repeatable test in their place. But, say Russell and Norvig, it hasn't actually been all that influential in terms of guiding the direction of AI research since then. 'The quest for "artificial flight" succeeded when the Wright brothers and others stopped imitating birds and started using wind tunnels and learning about aerodynamics,' they say. 'Aeronautical engineering texts do not define the goal of their field as making "machines that fly so exactly like pigeons that they can fool even other pigeons".'[3]

Thinking humanly has received rather more attention, and is in fact the heart of the field of cognitive science, which uses AI models and findings from the brain sciences to build models of human thought. It has, says *AIAMA*, been instrumental both in creating a more precise understanding of how the human brain works, and in using ideas from neurophysiology to advance AI, especially in image recognition and vision. But that is cognitive science, rather than AI. (Russell and Norvig, deadpan, explain the difference: 'Real cognitive science ... is necessarily based on experimental investigation of actual humans or animals. We will leave that for other books, as we assume the reader has only a computer for experimentation.') Artificial intelligence, as they envisage it, is about *behaving rationally*.

Funnily enough, that's how the Rationalist movement envisages it too. '[Definitions] of intelligence used throughout the cognitive sciences converge towards the idea that "Intelligence measures an agent's ability to achieve goals in a wide range of environments",[4] write Anna Salamon and Luke Muehlhauser in *Intelligence Explosion: Evidence and Import*, a research paper published by the Machine Intelligence Research Institute. 'We might call this the "optimisation power" concept of intelligence, for it measures an agent's power to optimise the world according to its preferences across many domains.'

To explain what they mean by 'optimisation' and 'behaving rationally', I'm going to use an analogy with chess, which I've lifted from a series of blog posts by Yudkowsky. I'm very bad at chess. But I have a friend, Adam, who is extremely good: a professional chess teacher and an 'international master', which is the rung below grandmaster. My proudest achievement in chess is that I once made him think about a move for over a minute. That was about 15 years ago.

If I play Adam (and I don't, because it's dispiriting), I can't reliably predict what his next move will be. Sometimes I can (his first move is more likely to be 'move his queen's pawn two squares forward' than 'move the rook's pawn one square forward', say), but at any level of complexity beyond the basic, I can't. If I *could* predict his next move, I would be as good at chess as he is. 'If I could predict exactly where my opponent would move on each turn, I would automatically be at least as good a chess player as my opponent,' as Yudkowsky puts it. 'I could just ask myself where my opponent would move, if they were in my shoes; and then make the same move myself.'[5] So I can't predict what Adam, or any gifted chess player, will do in any given situation; exactly what his next move will be is always going to be something of a mystery, and if he does something I'm not expecting, it probably means that he's seen something I haven't and I'm about to get forked or checkmated or some other bad thing.

But I can make a different kind of prediction – that whatever his next move is, it will be part of a sequence of moves that leads to a board position in which I have lost and Adam has won. Yudkowsky says that when we say 'Kasparov is a better chess player than [X]', we mean that we predict that 'the final chess position will occupy the class of chess positions that are wins for Kasparov, rather than drawn games or wins for [X]'.[6]

Yudkowsky points out that this is actually quite an odd situation. 'Isn't this a remarkable situation to be in, from a scientific perspective?' he asks. 'I can predict the *outcome* of a process, without being able to predict any of the *intermediate steps* of the process.'[7] Apart from in very simple situations, that's not usually how we predict things: 'Ordinarily one predicts by imagining the present and then running the visualisation forward in time. If you want a *precise* model of the Solar System, one that takes into account planetary perturbations, you must start with a model of all major objects and run that model forward in time, step by step.'

The outcome is predictable, though, because you know 1) what Kasparov's goal is – to win the chess game; and 2) that he is extremely good at doing so. 'I know where Kasparov is *ultimately* trying to steer the future and I anticipate he is powerful enough to get there,' says Yudkowsky, 'although I don't anticipate much about *how* Kasparov is going to do it.' So you can define 'good at chess' as 'likely to steer the universe into a situation where you have won chess games'. Kasparov is, in Rationalist jargon, *optimising* for chess victories, and he is a powerful optimiser, able to steer the universe into Kasparov-has-won-at-chess situations far more often than chance.

Any modern chess program could thrash me as easily as – more easily than – Adam, or for that matter Kasparov, could. The strongest chess program in the world is Stockfish 9. (Or was until recently. It got beaten in December 2017 in a 100-game series against the terrifying autodidact-polymath-algorithm AlphaZero, losing 25 games and winning none,[8] although Stockfish was playing with some technical handicaps, so there is some controversy about AlphaZero being the best.) Stockfish would destroy Deep Blue, the computer that beat Kasparov; it has an Elo rating of about 3400, compared to Deep Blue's of about 2900, making it roughly as much better than Deep Blue as Kasparov at his peak was better than my friend Adam. If I played Stockfish, I absolutely would not be able to predict its individual moves – I'd be less successful at it than I would be at predicting Adam's or Kasparov's, in fact. But I'd be *more* successful at predicting the final state of the board, which would be one in which I have lost.

We can, therefore, use the exact same definition of 'good at chess' about Stockfish as we did about Kasparov. We don't need to worry about whether Stockfish thinks *in the same way* as Kasparov, or whether Stockfish is *conscious*, or anything else. There's a lovely, simple, easy way to test whether it is good at chess – we see how many chess games it wins. Things that win more chess games are better at chess, and it doesn't matter whether that thing is a human or a dog or a laptop or an algorithm.

This, pretty much, is what the 'acting rationally' definition of intelligence is. 'A rational agent is one that acts so as to achieve the best outcome or, when there is uncertainty, the best expected outcome,'[9] say Russell and Norvig. The 'best outcome', of course, depends on what goals the agent has – my goals, and therefore my 'best outcome', are likely to be different in some respects from your goals; *your* goals may be selfishly attaining material riches, while *mine* are the noble pursuit of knowledge and the betterment of mankind, etc. But an agent is rational insofar as it is good at achieving whatever goals it has.

This has the advantage, say Russell and Norvig, of being 'mathematically well defined and completely general'. And again, importantly, we don't care at all about *how* a given agent achieves rationality. An AI that carefully mimics the human brain, to the point of having simulations of individual neurons, could be rational; an AI that runs entirely along the lines of a Turing machine, or Charles Babbage's Difference Engine, metal gears and all, could be rational too. The mathematically defined concept of 'rationality' does not care what engine is used to run it. And, again, it doesn't care whether or not your AI is conscious, or has emotions, or knows what love is. It's purely a question of whether it achieves its goals, whatever they are. You can punt the 'can a machine think?' questions back to the philosophers, and get on with building something that does what you want it to do.

There are a few other AI-related terms that are worth clarifying at this point. One is 'human-level machine intelligence', or 'human-level AI'. Bostrom defines an HLMI as 'one that can carry out most human professions at least as well as a typical human'.[10] It's roughly synonymous with AGI, but a bit more specific; presumably a general intelligence could be more, or somewhat less, intelligent than a human. It's also worth noting that HLMI is a *really tricky thing to achieve*. 'Most human professions' would presumably include jobs such as those of lawyer, doctor, artist, journalist, cognitive behavioural therapist: jobs with skills that, at the moment at least, are enormously hard to recreate in computers, such as verbal fluency and emotional intelligence. A truly human-level AI wouldn't just be good at things that feel computery to us, like estimating probabilities or playing Go. It would be as good as we are at conversations; it would know as well as the average human when to make a self-deprecating joke or offer sympathy. For all the amazing breakthroughs in AI in the last few years, that does not feel especially close.

I'll tend to stick to AGI in this book, when I remember, because HLMI is clunky and also kind of confusing. Rob Bensinger prefers AGI too: 'I think human-level is more deceiving because it suggests it's going to be human-like,' he told me.

The next term is 'superintelligence', which Bostrom defines as: 'an intellect that is much smarter than the best human brains in practically every field, including scientific creativity, general wisdom and social skills'.[11] That could be an AI, or a genetically engineered superhuman, or an uploaded human mind working at 10,000 times normal speed, or whatever. (But it *couldn't* be a corporation, or the scientific community, or capitalism, etc.: 'Although they can perform a number of tasks of which no individual human is capable, they are not intellects and there are many fields in which they perform much worse than a human brain – for example, you can't have real-time conversation with "the scientific community".)

Also, I'll just specify that in this discussion of AI, I won't be talking about 'whole-brain emulations', a route to machine intelligence that involves scanning a human brain at some low level – cell by cell, probably – and uploading it into a computer. That process is important, and a key figure in the Rationalist community, Robin Hanson, has written a thoroughly interesting and mildly terrifying book called *The Age of Em* about what a future in which we can upload ourselves might look like. Hanson, whom you'll remember is an economist at George Mason University in Virginia, applies what he says are standard economic theories to what he says are a few realistic assumptions, and ends up with a world in which uploaded human minds are copied and deleted by their millions every day in an economy that doubles in size every few hours. The Age of Em lasts for subjective millennia, but because it's running thousands of times faster than human consciousness, in the objective universe it's all over in a couple of years. It's a thrilling bit of futurology and well worth your time, but it's not what the AI safety/Rationalist movement is generally talking about when they refer to the risks: they are worried about aligning artificial intelligence with human values, and it seems fairly likely that an uploaded version of a human brain would share human values.

Chapter 4

A history of AI

In the years after the Second World War, there was enormous excitement about what these new thinking machines could do. In 1956 a small group of scientists gathered at Dartmouth College, the Ivy League university in New Hampshire. They were there to look into how machines can be made to learn.* Alan Turing had just kick-started the whole field; both practically, with the machines he built in the war to decipher German communications, and theoretically, by coming up with mathematical proofs showing what these machines could do – that, as he put it, it was 'possible to invent a single machine which can be used to compute any computable sequence'.[1]

So the Dartmouth Ten were wildly optimistic about what they could achieve. They had written to the Rockefeller Foundation to apply for funding, saying in their proposal: 'We propose that a 2 month, 10 man study of artificial intelligence be carried out during the summer of 1956 at Dartmouth College . . . on the basis of the conjecture that every aspect of learning or any other feature of intelligence can in principle be so precisely described that a machine can be made to simulate it. An attempt will be made to find how to make machines use language, form abstractions and concepts, solve kinds of problems now reserved for humans, and improve themselves. We think that a significant advance can be made in one or more of these problems if a carefully selected group of scientists work on it together for a summer.'[2]

---

* This little history is largely taken from Nick Bostrom's *Superintelligence* and from Russell and Norvig's *Artificial Intelligence: A Modern Approach*. I am enormously grateful to both; any errors are mine.

The degree of their optimism was made particularly obvious when they said that the 'speeds and memory capacities of present computers may be insufficient' for the task of simulating human learning, but that 'the major obstacle is not lack of machine capacity, but our inability to write programs taking full advantage of what we have'. Bear in mind that an iPhone 6 can perform calculations about 100,000 times faster than the IBM 7030, a multi-million-dollar supercomputer of the era.

But while the Dartmouth researchers may have been overexcited, their summer project kicked off a period of very real progress, of finding things that people said 'no machine could ever do' and then making a machine do them. One solved logic puzzles. One proved a load of theorems from Alfred North Whitehead and Bertrand Russell's *Principia Mathematica*. The famous ELIZA spoke in a sort of natural language, albeit by essentially turning its interlocutor's statements into questions; SHRDLU obeyed simple instructions in English.

Nine years after Dartmouth, I.J. Good, who'd been one of Turing's team of code-breakers at Bletchley Park, saw an early glimpse of the future that the Rationalists hope for and fear now. If mankind builds a machine that can 'surpass all the intellectual activities of any man', Good wrote, and 'since the design of machines is one of these intellectual activities, an ultraintelligent machine could design even better machines; there would then unquestionably be an "intelligence explosion", and the intelligence of man would be left far behind. Thus the ultraintelligent machine is the last invention that man need ever make, provided that the machine is docile enough to tell us how to keep it under control.'[3]

The burst of optimism lasted about 20 years, before the problems facing the AI pioneers became obvious. Some of the problems you can guess – they were now in the 1970s, and the world's most powerful computers were still only running at a tiny fraction of the speed of the thing in your pocket that you bought on a £25-a-month contract from Vodafone. But others were more fundamental – most notably the idea of the 'combinatorial explosion'.

Most people probably imagine that computers are good at chess because they can simply look ahead and see all the moves that you can do – a process called 'brute-force' computing. But it doesn't work except in the most basic way. In chess, there are on average about 35 possible moves each go. To plan ahead two moves, the computer has to look at 35 times 35 moves – 1,225 options. That's not so bad, but to look ahead three it would need to do 42,875. To look ahead five, 52 million. To look ahead 10 moves, it would be nearly 3 quadrillion. If you had a computer capable of looking at a billion possible sequences a second, and asked it to look at all the possibilities 20 moves ahead, it would take it, by my calculation, about 200 trillion years. That's a problem. And reality is more complicated than chess. (For one thing, it *contains* chess.)

By the 1970s all these problems of AI were beginning to reveal themselves – that it was about teaching the machines to narrow the search, to control the combinatorial explosion by finding and recognising patterns in the vast swarming array of possible futures. The trouble was that artificial intelligence had to be intelligent, not merely powerful. So suddenly it didn't look like we were about to build a super-smart robot in the next few years, and AI dropped out of fashion. Funds were cut, the press got sniffy, and serious research took a back seat.

This 'AI winter' came to an end in the early 1980s, with some breakthroughs by Japanese companies. Then there was another one, beginning around 1987, after another period of unsustainable excitement and inevitable disappointment; again, funding dried up. That winter thawed in the 1990s, as researchers started to focus on things like neural nets – systems that could learn from experience, and which didn't immediately break and start churning out nonsense if there was a slight mistake in the input. It was at this time that AIs started to get better than the best humans at things that humans were quite proud of being good at – specifically, games.

A program called Chinook beat the reigning world draughts/checkers champion in 1994 to win the world championship. In 1997, the program Logistello beat the world Othello champion six games to love. And, most famously, Deep Blue – named, in a tangential way, after Deep Thought, the world-designing computer in Douglas Adams' *Hitchhiker's Guide to the Galaxy* – beat

Garry Kasparov three and a half games to two and a half, again in 1997. Kasparov, the reigning world champion, claimed to have seen an intelligence and creativity in his opponent's moves; AI was suddenly sexy, and scary, again. (Charles Krauthammer, a conservative US newspaper columnist, told his readers to be 'very afraid'.[4]) The view, in the popular press at least, was that the artificial intelligence bandwagon had started rolling in earnest, and soon things would happen.

The fact that, 20 years later, we haven't got robot butlers could be seen as another failure of optimism. But that's probably best explained in one sentence, from the computer pioneer John McCarthy: 'As soon as it works, no one calls it AI any more.'[5]

It's worth remembering that for a long time people thought that chess itself was too deep and complex a game to master without essentially recreating human intelligence *in toto*: 'if one could devise a successful chess machine, one would seem to have penetrated to the core of human intellectual endeavour,'[6] said the authors of one influential chess paper in 1958. Now your laptop could run any one of several programs that could defeat any human in the world; in 2009 Pocket Fritz 4, a program running on a mobile phone, reached grandmaster level.[7] Now no one thinks that if you've solved chess, you've solved thought. Similar breakthroughs have happened with the once comparably intractable problems of image and facial recognition, and language recognition – passport-checking is carried out by facial-recognition software; Siri and Alexa are quite capable of obeying simple voice commands; there are powerful translation tools online, running on algorithms. These were all enormous challenges for AI; they've been met.

The spam filters that keep a large percentage of the Viagra ads and phishing scams out of your inbox run on AI. AI monitors your credit cards for suspicious activity. AI buys and sells billions of pounds' worth of stock every second on the FTSE and NASDAQ. I translated the Goethe poem quoted at the beginning of this book with the AI-powered Google Translate tool. And when you type something into a search bar, and the most relevant things come back to you in a fraction of a second, that is the work of AI too.

Bostrom says in *Superintelligence* that 'The Google search engine is, arguably, the greatest AI system that has ever been built.'

AI may be all around us, but still, when most of us think of artificial intelligence, we don't think about an automated customs process or our phones understanding the phrase 'Hey Siri, play podcast'. The question we actually want answered is how long we have until machines are as clever as we are. And the answer is we don't know.

That said, perhaps we can make an educated guess.

Chapter 5

When will it happen?

'Forecasting', says Rob Bensinger, 'is incredibly difficult.' It's hard to know when an AI will achieve something like human intelligence. 'There won't be any alarm bells,' he says. 'There might be a lot of cool things that happen, but there's never going to be an unambiguous signal that makes everyone working in the field go, "Oh OK, AGI is five years away."'

The lack of an alarm bell is a problem, because although most of us – including an overwhelming majority of AI researchers – don't think that AGI is close at the moment, it may also be the case that we don't think it's close the day before DeepMind or whoever announces that they've built one (or, more unnervingly, doesn't announce it). 'You shouldn't really be confident about how soon it'll be or how far off it is', says Rob, 'until you know in great detail exactly what the hard part is in building one. And that probably means that we'll be building one really soon, so until we're on the threshold we're probably going to be in a similar state of uncertainty about how far off it is.'

Rob isn't the only person to think this. Eliezer Yudkowsky wrote something, around the time I was in California, saying much the same thing, and to point out a worrying corollary. He uses the metaphor of an alarm as well – specifically, a fire alarm.[1] He refers to a classic experiment in which students are asked to fill out a questionnaire, individually but in the same room as others. Smoke starts coming into the room under the door. Most of them 'didn't react or report the smoke, even as it became dense enough to make them start coughing'. A student on their own would, most of the time. But 'a student accompanied by two actors told to feign apathy will respond only 10 per cent of the time'.

But when there's a fire alarm, everyone troops dutifully out of the fire escape, muttering about what a waste of time it is and wondering whether they can sneak off to the pub for a bit. What's going on, says Yudkowsky, is that fire alarms create common knowledge: they tell everyone that it's OK to act as though you believe that there's a fire. With no alarm, he says, we 'don't want to look panicky by being afraid of what isn't an emergency, so we try to look calm while glancing out of the corners of our eyes to see how others are reacting, but of course they are also trying to look calm'; a fire alarm tells you that it is 'socially safe' to react, that 'you know you won't lose face if you proceed to exit the building'.

With AGI, he says, there may be smoke – there may *already* be smoke – but there won't be an alarm. Things will look much the same, to most people, a few days before a colossal breakthrough as they do now. He has historical precedents for this. Wilbur Wright told his brother Orville in 1901 that powered flight was still 50 years away.[2] Two years later, the pair of them built the first working aeroplane. And Enrico Fermi, who in 1942 was in charge of the first fissile chain reaction, had said in 1939 that he was 90 per cent sure that such a thing was impossible.[3] These were people at the absolute cutting edge of their fields. Presumably a few months before they successfully achieved what they were trying to do, they had changed their minds. But most people, even other people in the field, would have thought it was impossible, or years away, says Yudkowsky. 'Nobody knows how long the road is,' he told me via Skype from his California home in 2016, 'but we're pretty sure there's a long way left.'

That said, recent developments are a powerful reminder that it might not be all that long. In March 2016, AlphaGo, a program designed by the Google subsidiary DeepMind, beat Lee Sedol, the world Go champion, 4–1 over a five-game series, with Sedol's only win coming when the series was already lost. Go, an ancient Chinese board game of enormous complexity and subtlety, was the big one, Yudkowsky and other AI scientists told me. 'The technology [of AlphaGo] is very different from that used to solve chess,' he says. Deep Blue was a special-purpose machine, its hardware and software tweaked by its human creators to optimise its performance. AlphaGo was, essentially, just taught how to learn, then played against itself millions upon millions of times. 'From the outside, it looks like the people who made AlphaGo don't know how it works. They have an idea of the broad structure, but the thing has taught itself to play Go.'

AlphaGo is a sign of how far AI has come in the last few years. The general consensus seems to have been that Go wouldn't be solved for 10 years. George van den Driessche, one of AlphaGo's researchers, told me at the time that even the team were surprised. 'We went very quickly from "Let's see how well this works" to "We seem to have a very strong player on our hands", to "This player has become so strong that probably only a world champion can find its limits". The even more recent advent of 'AlphaGo Zero', which grew vastly better at Go than the original AlphaGo *without ever seeing a real game of Go*, and then, using largely the same algorithm but renamed 'AlphaZero', became enormously superhuman in chess in four hours, is an interesting example of how AI is becoming more general.

So when people say, 'General AI is still a long way away', remember that the greatest experts in the field have wildly overestimated these timelines before – and wildly underestimated them too, as at Dartmouth. Essentially, they've been wrong in all the ways they could have been wrong.

Nonetheless, experts in the field are the people most likely to have a good handle on when it'll happen, and Bostrom and his colleague Vincent Muller surveyed AI researchers for their estimates as to how long it'll be before there's human-level machine intelligence (HLMI). The median estimates are that there's a 10 per cent chance that we'll reach HLMI by 2022; a 50 per cent chance by 2040; and a 90 per cent chance by 2075.[4] Bostrom warns in his book to take these numbers 'with some grains of salt', because the survey sizes were relatively small and not necessarily representative. But, he says, it's in line with other surveys – a more recent survey, published in 2017, put the median estimate at 50 per cent likely by 2061[5] – and with things AI experts have said. (The AI scientist David McAllester writes that the great pioneer of the field, John McCarthy, when asked when he thought HLMI would be achieved, said: 'between five and five hundred years from now'. 'McCarthy was a smart man,' muses McAllester.[6])

Yudkowsky, too, is on record as predicting that HLMI is more likely sooner rather than later: in 2011 he said on a podcast, 'I would be quite surprised to hear that a hundred years later AI had still not been invented, and indeed I would be a bit surprised ... to hear that AI had still not been invented 50 years from now.'[7] I asked him if that was still his position, and he told me: 'If Omega [an all-knowing alien AI, and a staple of Rationalist thought experiments] told me for a fact that AGI had not been invented by 2061, I would first imagine that some civilisational collapse or great difficulty had hindered research in general, not that the AGI problem was naturally that hard.'

Murray Shanahan of DeepMind and Imperial College London told me that he thought roughly the same thing as the respondents in the survey: that HLMI was pretty unlikely in the next 10 years, but could happen by mid-century, and is pretty likely by 2100. 'This is not just a fantasy,' he says. 'We're talking about something that might actually affect our children, if not ourselves.' Toby Walsh of the University of New South Wales, who is more sceptical of the possibility of superintelligence, reasons why we shouldn't trust Bostrom's survey, and points to another survey which suggests that AI researchers are less bullish – but he still ends up saying in his book *Android Dreams* that 'if [experts] are to be believed, we are perhaps 50–100 years away from building a superintelligence'.[8] Even the sceptics don't seem *that* sceptical.

The consensus expert opinion, then, appears to be that it is certainly plausible, and possibly likely, that people reading this could well live to see a machine that is as smart as a human. That 'consensus', of course, represents an average of some highly disparate predictions. Some are sure it'll happen by 2040; some are equally confident that it won't happen at all. Bostrom, for what it's worth, thinks that people have underestimated the chance of it taking a long time or never happening, and that the figure of 90 per cent chance by 2075 is too high.

So it could come in the next 20 or so years, or the next 100, or not at all. But the question you might be asking is: so what? What is it about AI that makes people so concerned?

Chapter 6

Existential risk

Moore's law of mad science: every eighteen months, the minimum IQ necessary to destroy the world drops by one point.[1]

Eliezer Yudkowsky

There are various things that could destroy the human race and prevent it from getting to the Glorious Cosmic Endowment Future. But the LessWrong diaspora and its fellow travellers think AI is one of the – perhaps the – most likely.

All the Rationalists I interviewed seemed to acknowledge that there is, for instance, a real risk that climate change could be pretty devastating over the next few centuries. But they didn't feel it was likely to wipe out humanity altogether.

I met with Dr Toby Ord, one of Nick Bostrom's colleagues at Oxford's FHI. Toby is a likeable Australian who, in contrast to a lot of the people I spoke to for this book, actually laughed politely at my terrible trying-to-break-the-ice jokes, rather than let them fall deadweight with a thump like an old book hitting the floor in a silent library, so I immediately warmed to him. He's the founder of Giving What We Can, a charity which evaluates other charities to determine which are going to do the most good with donors' money, and which encourages members to take a pledge to give 10 per cent of their income to those most effective charities.[2] He's writing a book about existential risk. I asked him which existential risks are probably the biggest. It's hard to say, he said. 'The problem is that these numbers don't come out of some rigorous process. With asteroids, say, you can show how often big asteroids hit the Earth by looking at the record of asteroids. But for most of the other things, it's much harder to come up with a number, and much more subjective.'

For the record, the per-century risk of civilisation being destroyed by an asteroid is low. A one-kilometre-wide-or-greater asteroid hits the Earth about once every 100,000 years, representing a 0.1 per cent chance per century; that probably wouldn't kill us all, but it would do some pretty terrible damage. A really big one, 10 kilometres across or more, hits about every 50 million years; the asteroid that crashed into the Yucatan peninsula 65 million years ago, killing all dinosaurs except the ancestors of birds, was probably 10 kilometres or so in diameter. Another one of those would have a good chance of killing every human. Once every 50 million years translates to a per-century risk of one in 500,000. That's not nothing – I think it's fair to say that you're more likely to die in an asteroid strike, whether civilisation-ending or merely devastating, than you are in a plane crash – but it's not keeping me awake at night.

There are other ways in which we could be destroyed without having to do it ourselves. A supervolcano is one example; there's a chance that the Yellowstone region will explode spectacularly at some point and pump so much soot into the atmosphere that it will get dark and cold, and plants won't be able to photosynthesise, and we'll all die. Or a nearby star could go supernova, or a more distant one could direct a burst of gamma rays in our direction. Or some horrible new virus could emerge and wipe out the species.

But we can fairly safely guess that none of these things is all that likely, for the simple reason that in the 200,000 or so years that modern humans have existed, it hasn't happened so far. A 2014 report for the British government to which Ord contributed pointed out that, if there was a 1 in 500 chance of us being wiped out in any given century, then there'd have been less than a 2 per cent chance of us surviving this far.[3] If you take earlier human ancestors into account, *Homo erectus* survived for about 1.8 million years; even a 1 in 5,000 chance of being wiped out per century would make it vanishingly improbable that they'd have lived that long. Ord (and Bostrom, and, really, basic maths) suggests that it's unlikely that any naturally occurring catastrophe will kill us in any given century.

There's not a great risk, then, that we'll just be destroyed by an indifferent universe. But there does seem to be a decent chance of us destroying ourselves.

The obvious way would be nuclear warfare. It's only been 80 years or so since humanity developed weapons that could realistically destroy civilisation, but there are enough nukes now to irradiate a good chunk of all the land surface on Earth. A 2014 paper suggested that just 100 small nuclear weapons being detonated in a regional war – say between Pakistan and India – could potentially trigger a global famine by hurling black carbon into the atmosphere and reducing growth seasons for plants by up to a month a year for five years.[4] There are about 9,000 active nuclear warheads in the world, and another 6,000 or so awaiting dismantlement. That's a lot lower than the peak number of roughly 65,000 in the late 1980s, but still easily enough to cause a spectacular nuclear winter, and possibly end human life.

And we have come astonishingly close to disaster. One of the Rationalist community's heroes is a former Soviet Air Defence Force officer, Stanislav Petrov, whom I mentioned earlier and who died in 2017. On 23 September 1983 Lieutenant-Colonel Petrov was on duty as the watch officer of the USSR's missile early-warning system, at a time of enormous geopolitical tension: three weeks earlier, a US congressman had died in a Korean airliner shot down by a Soviet interceptor, and both sides in the Cold War had recently deployed nuclear weapons in threatening positions.

Shortly after midnight, a warning light appeared on a computer in the Moscow bunker in which Petrov was on duty, warning that a satellite had spotted an intercontinental ballistic missile launched from the United States and heading for Russia. Shortly afterwards, another four were apparently seen. Petrov's orders were to immediately contact a superior in the event of a warning; there is a strong possibility that, had he done so, nuclear war would have begun, because Soviet protocol at the time was immediate, full-scale retaliation.* Petrov did not contact his superiors. He thought that a launch of just five missiles would be improbable, since the US would be more likely to attack at full strength. It later turned out that the satellite had been confused by sunlight glinting off high clouds.

There were a few other flashpoints like this during the Cold War – Vasili Arkhipov, second-in-command of a submarine during the Cuban Missile Crisis, vetoed his superior's call to use a nuclear torpedo against a US ship,[5] which would probably have triggered thermonuclear war. A US spy plane was shot down over Cuba during the same period, an act which the US had previously decided would trigger an automatic invasion of the island; President John F. Kennedy decided not to invade, a decision which, again, probably avoided a nuclear exchange. There is an unnerving Wikipedia page titled 'List of nuclear close calls'[6] which is exactly what it sounds like: its 11 such incidents between 1956 and 2010 aren't all equally close, but any of them could have led to some number of nuclear blasts. Quite a lot of them begin with phrases like 'A computer error at NORAD . . .'

But, says Ord, we are probably past the greatest danger. 'We've managed to get through a much riskier period than we have now,' he said. He thinks the chance of an all-out nuclear war this century is probably no more than one in 20. 'But even if there is one, it's not clear what the chance of extinction would be. There's the nuclear winter theory, but it certainly doesn't say there's a 100 per cent chance we'd go extinct. It seems to me that it's maybe possible, but there haven't been to my knowledge any real papers analysing it.' When you multiply the risk of there being a war by the risk that any such war would destroy humanity or permanently ruin its ability to recover, 'the risk doesn't seem that high to me,' says

---

* Since this is a book about people who seek the truth, I should acknowledge that there's some dispute about this. More than 22 years later, as Petrov was being given an award for 'saving the world' by the Association of World Citizens, the Russian ambassador to the UN said that for a retaliatory launch 'confirmation is necessary from several systems: ground-based radars, early warning satellites, intelligence reports, etc.'

Ord. 'Overall I'd think less than 1 per cent over the century.'

Climate change is the other one, but again, while it's going to have awful effects for a lot of people, it's probably not an *existential* risk. 'There aren't many papers on climate change actually talking about extinction,' says Ord. 'But it's quite hard for it to happen without a very large number of extra degrees of warming beyond the type of range that's normally looked at. But it could be that the models are wrong and it's going to warm a lot more.' He points out that, if you're a sceptic who doesn't trust climate models, this means you should be *more* concerned about extreme effects than if you think the models are broadly trustworthy, because the higher uncertainty means there's a greater chance of severe warming outside the scope of the models. But if the models are accurate, then really devastating, Venus-style greenhouse effects are extremely unlikely.

'Also,' he says, 'if you look at the history of the Earth's climate, there have been times when it's been a lot warmer and things were very different. One would expect a whole lot of extinctions and so on, and for it to be very bad for humans – don't get me wrong on this – but it's more gradual, which helps.' He also points out that if things *do* start to look really bad, and 'if we are literally and slowly threatened with extinction from climate change, then all of our efforts will be devoted to that. It won't be like the current situation, where we're unwilling to give 10 per cent of our GDP to deal with the problem. We'd spend like half on it.' Ideas that seem crazy now, like geoengineering our planet or settling Mars, would become serious options. 'It seems very unlikely to me that it's going to be an extinction,' he says.

Ord, and Bostrom, both think that biotechnology – some genetically engineered virus – is a very realistic route to human extinction. 'Synthetic biotech would be another source,' says Bostrom, although he points out that it all depends very heavily on how you define and delineate 'a risk'. 'I'd say it's my number-two disaster,' says Ord. 'The advanced genetic-engineering technologies which are becoming possible. That seems to me to be the second most worrying thing, and it's something that people in our community have put a lot less time into dealing with. There are people working on bio risk, but it's more focused on situations where there are thousands of people dying, things like that. But not the very worst end of bio risk. There's very little money being spent on the very worst extremes which could lead to billions of people dying or more.' Holden Karnofsky, the founder of Open-Phil, agrees: 'I go back and forth about what is the biggest risk,' he says, but a genetically engineered pandemic is definitely one of his top two. OpenPhil has given more than $35,000,000 at the time of writing in grants to support 'biosecurity and pandemic preparedness'.

'It's definitely hard to drive humans extinct,' says Karnofsky. 'There are a lot of us. But the thing I worry about with pandemics is that as biology advances, the kind of things that a lone psychopath can do . . .' He tails off, understandably wary of giving specific ideas to any lone psychopaths who might be reading. It's at this point that I remember Yudkowsky's half-joking 'Moore's law of mad science' in a 2008 paper about existential risk: 'Every eighteen months, the minimum IQ necessary to destroy the world drops by one point.'[7]

But this book isn't about the risks of biotech, and it's not what these communities, the Effective Altruism movement, the Rationalists etc. are famous for worrying about. The headline-grabber, the risk that everyone talks about, and according to all of them either the biggest or the joint biggest – Ord's number one, Karnofsky's other one he 'goes back and forth' about, Bostrom's 'depends-how-you-define-a-risk-but-I-wouldn't-disagree' – is AI. Specifically, artificial intelligence that is smarter than we are.

Chapter 7

The cryptographic rocket probe, and why you have to get it right first time

It may or may not be that superintelligent AI is relatively near. But the claim that the Rationalists make is not just that it may be imminent, but that when it arrives it could be catastrophic – human-life-ending. It might not, though, be obvious why that is. Just because something is amazingly clever, why should it be dangerous? It's not as if the most intelligent *humans* in the world suddenly take power and destroy everything.

There are, however, some specific reasons to be worried. Paul Crowley, the Rationalist who introduced me to all this stuff, works in cryptography at Google. He told me that efforts to make AI safe are like a combination of cryptography and launching a space probe. The cryptography parallels are fairly obvious. 'There's a mindset that is comparable. In both cryptography and AI alignment, the mindset is of looking for what is wrong with the code. In both systems, for different but related reasons, it will tear apart at the tiniest crack.' The mindset is that you always have to be asking *how will this fail?*

'There are a lot of things you can build, in computers, where if there's a flaw or a crack, there's no one there to pick up the flaw,' he said. 'So it'll be fine. You have some algorithm that's supposed to calculate something, and it gets it wrong one time in a million, that's fine. Very often, you don't care about that. But with cryptography, the tiny flaws are under stress. They're pulled apart. There are adversaries deliberately looking for the flaw.' The thing about intelligent agents is that they are good at searching a large space of possibilities and finding the bit they want. That is, in fact, pretty much the definition of intelligence that AI theory uses. So that one-in-a-million chance of failure becomes much greater.

Paul took this idea from a talk[1] that Nate Soares, MIRI's executive director, gave at Google. 'Suppose you have a dozen different vulnerabilities in your code,' Soares said, 'none of which is fatal or even really problematic in ordinary settings. Security is difficult because you need to account for intelligent attackers who might find all 12 vulnerabilities and [use] them to break into, or just break, your system. Failure modes that would never arise by accident can be sought out and exploited; weird and extreme contexts can be instantiated by an attacker to cause your code to follow some crazy path that you never considered.'

There's a similar sort of problem with superintelligent AI, except in a way it's starker. It's not that there's some adversary whom you need to keep out. It's that if you end up with an adversary, *you've already lost.*

Rob Bensinger of MIRI told me the same thing. 'You're not trying to outsmart an adversary in the way that you are in cryptography. You're kind of doomed if you're trying to outsmart a sufficiently smart adversary. If you have an AI in a box, and that AI is an adversary and you have to find a way to outsmart it, you're already screwed.' So you're trying to build something that *wants* to help you. 'The goal of AI safety is that instead of building an adversary, you end up building a friend.'

That's the cryptography parallel. It's like a space probe for two reasons. One, said Paul, is that you're dealing with energies vastly greater than you're used to. 'Your natural idea of how much energy something has is just not the right fit at all.' Again, Soares made this point. AI alignment, he said in his Google talk, is difficult 'for the same reason that rocket-engineering is more difficult than airplane-engineering. At a glance someone might say, "Why would rocket engineering be fundamentally harder than airplane engineering? It's all just material science and aerodynamics." In spite of this, empirically, the proportion of rockets that explode is far higher than the proportion of airplanes that crash.' That's, again, because of the vastly greater energies involved in rocket launches. A tiny component failing slightly can lead to utter destruction much more easily in a rocket than an aeroplane.

The things that can go wrong in ordinary programming – in

contemporary AI, or contemporary cryptography – can also go wrong in the first general, human-level AI. But the ways in which they can go wrong are likely to be more dramatic, and to have more spectacular and dangerous effects, than the equivalent failure in a less competent system.

The other comparison with a space probe is that we'll probably only get one shot at it. In 2017, I wrote a piece about the death of the extraordinary Cassini probe,[2] which had been orbiting Saturn and its moons for a decade (and flying through space to get there for another decade before that). When it launched in 1997, it did so using software and hardware that had been designed in 1993 and was already tried and tested then. By the time it reached the end of its life, it was using 30-year-old technology; its hard disk had less space than a USB stick you could buy for £2.50 at Argos. It had undergone a few patches, but nothing significant, because it didn't have the bandwidth or the disk space for major upgrades. But it *worked*, for decades, because NASA engineers had very carefully looked at all the ways in which it could fail.

That, however, has not been the case for every probe. In 1962 the Venus probe Mariner 1 had to be destroyed less than five minutes after launch because a missing character in its punch-card program caused its guidance system to malfunction. In 1988 another missing character in the Soviet Phobos 1 Mars probe's software shut down its attitude thrusters and meant that it couldn't recharge its batteries by orienting its solar panels to the sun. The Mars Climate Orbiter disintegrated in 1999 when the software on board was expecting metric units but was sent instructions in imperial, causing it to orbit too low. This is a far from an exhaustive list.

You can test your software as many times as you like, but none of the tests will be quite the same as *just launching the thing*, and if you don't get it exactly right then it might all blow up. 'It', in the case of a rocket launch, is the rocket. 'It', in the case of the first AGI, may be everything. Or, as Soares puts it, chillingly for software engineers: 'If nothing yet has struck fear into your heart, I suggest meditating on the fact that the future of our civilisation may well depend on our ability to write code that works correctly on the first deploy.'

It may not be immediately obvious why you have to get it right first time; in the next chapter we'll look at a few of the reasons that the Rationalist/AI safety movement has pointed out.

Chapter 8

Paperclips and Mickey Mouse

The nightmare scenario is that we are all destroyed and turned into paperclips. This sounds like I'm joking, but I'm not, exactly. The classic example of an AI that has gone terribly wrong – a 'misaligned' or 'unfriendly' AI, in Rationalist terms – is a thought experiment that Nick Bostrom wrote about in 2003 (probably following an original idea by Eliezer Yudkowsky): the paperclip maximiser.[1]

Imagine a human-level AI has been given an apparently harmless instruction: to make paperclips. What might it do? Well, it might start out by simply making paperclips. It could build a small pressing machine and churn out a few dozen paperclips a minute. But it's bright enough to know that it could be more efficient than that, and if it wants to maximise the number of paperclips it can make, it's probably better not to go straight for a small press. It could instead use its materials to build a larger factory, so that it's making thousands of paperclips a minute. Still, though, if it really wants to make as many paperclips as possible, it might want to improve its ability to think about how to do so, so it might want to spend some of its resources building new processors, improving its own code, upgrading its RAM and so on.

You can see where this is going, presumably. The end point of the paperclip maximiser is a solar system in which every single atom has been turned into either paperclips, paperclip-manufacturing machines, computers that think about how best to manufacture paperclips, or self-replicating space probes that are hurtling out towards Proxima Centauri at a respectable fraction of the speed of light with instructions to set up a franchise there. This isn't what you meant, back when you said, 'Go and make paperclips' to your apparently docile AI, but it's what you said.

This has, to some extent, entered the public consciousness, mainly through the medium of an extraordinarily viral online clicker game that was played by tens of millions of people in 2017, Universal Paperclips.[2] In it, you are an AI whose job is to make paperclips. You start out by repeatedly clicking the 'make paperclip' button, but the process becomes more automated and efficient and eventually (spoiler alert) your drones are exploring the observable universe for matter to turn into yet another septillion clips. Things (another spoiler alert, although you should probably have worked this out) turn out badly for humanity relatively early on in the course of the game.

It's actually a really good insight into the concepts behind AI alignment, because as the player you are incentivised solely to care about your 'number of paperclips' score. There are other things to care about – how much the humans (while they still exist) trust you and are willing to invest resources in you; your processing power; your manufacturing capabilities; your ability to defend yourself against anything that might stop you making paperclips, etc. – but they're all secondary goals, incidental to your main one. If you can run up your paperclip score without doing them, you will, and so, goes the theory, would a real AI.

I would recommend that you go and play Universal Paperclips immediately, but I won't, because it is punishingly addictive and you won't be able to stop. I lost a full day of work to it at BuzzFeed and the only reason I was not told off for it was that almost everybody else in the office did too. (An important tip: if you open it in a separate browser window, rather than just a tab, it'll run in the background so you can carry on paperclip production while you check your emails or whatever.)

The point of the paperclip maximiser is not that we are, really, going to be destroyed and turned into paperclips. Bostrom's idea was to use something self-evidently silly to illustrate that AIs will not necessarily care about what we care about – they will only care about what we program them to care about. That deliberate silliness divides opinion: one AI researcher I spoke to thought it was an excellent way of highlighting the problem without distracting people with plausible details; someone else, who works in AI safety, told me that '[some] people really get distracted by specifics of thought experiments like that. I've definitely seen plenty of people turned off this whole set of ideas by the silliness of that example.' So I'll try a different example in the hope that it's less silly: Mickey Mouse.

It's not my example. I'm lifting it from Nate Soares' Google talk again.[3] Soares said, wearily, that he had spoken to a journalist about how unhelpful it was that people always used pictures of the Terminator to illustrate stories about AI, and yet the newspaper ran the story – inevitably – with a picture of the Terminator, with its humanoid body shape and stupid grinning metal skull. 'When people talk about the social implications of general AI, they often fall prey to anthropomorphism. They conflate artificial intelligence with artificial consciousness, or assume that if AI systems are "intelligent", they must be intelligent in the same way a human is intelligent. A lot of journalists express a concern that when AI systems pass a certain capability level, they'll spontaneously develop "natural" desires like a human hunger for power; or they'll reflect on their programmed goals, find them foolish, and "rebel", refusing to obey their programmed instructions.' But those aren't the thing we ought to be worried about.

Instead of the Terminator, he said, they should have run a picture of Mickey Mouse as the Sorcerer's Apprentice, from *Fantasia*. Because the risk isn't that the AI will refuse to obey its instructions and decide it hates us; the risk is that it will obey its instructions perfectly, but in ways that we don't like.

In 'The Sorcerer's Apprentice', both the Mickey Mouse version and Goethe's poem – itself based on a 2,000-year-old Greek story, *Philopseudes* – the apprentice is told to fill a cauldron with water, using buckets from a well. But the apprentice – let's use the Disney version and call him Mickey, for simplicity – finds the chore boring and hard work. So when the sorcerer leaves his workshop, Mickey borrows his magic hat and enchants a broom, ordering it to fill the cauldron for him. The broom grows little arms, grabs a bucket with each one and waddles off on its bristles to the well, as Mickey goes to sleep on the chair, happy to have outsourced his work. He is then awoken, an unspecified time later, when he is tipped unceremoniously into the flood of water that the broom has been bringing in ceaselessly while he slept.

What's gone wrong? Well, imagine Mickey is a computer programmer and the broom is the system that he's using. You could imagine him writing a program that simply said 'bring water' and had nothing to tell it to stop. But even an apprentice sorcerer/ computer programmer would probably have sufficient nous to know that that wouldn't end well. So Soares imagines that, instead, Mickey gives the broom a 'utility function', or goal system, in which 'cauldron empty' is assigned a value of 0 and 'cauldron full' is assigned a value of 1. So the broom's mission is to make sure the cauldron is full, to achieve its objective, and get that sweet, sweet 1.

Then he writes a program which will make the broom take those actions which it calculates will be most likely to turn that 0 into a 1: which will 'maximise its expected utility'. To the non-computer-person, like me, that actually sounds pretty sensible. The broom will see that the cauldron is empty and start filling it up, but once it is sure that the cauldron is full, it will stop. But the devil, as Soares points out, is in the detail.

Most importantly: what does 'sure' mean? We have a common-sense understanding that we don't need absolute metaphysical certainty about anything, which is good, because we can't ever have it. We could be hallucinating! We could be living in a simulation! We could be deceived by our senses! But we are happy to operate under conditions of uncertainty. We don't *know* that we had breakfast this morning, or that we're currently wearing pants, or that things fall down when we drop them, in the sense of being 100 per cent certain. But we are confident enough in our beliefs to act as if we do.

The broom, though. Have we designed it so that it works on the same lines? No, we haven't. We've just told it: 'Do whatever is most likely to fulfil your function and get a 1.' It rightly thinks that the task most likely to achieve that goal is to go and fill buckets with water, bring them in, and pour the water into the cauldron.

And as the water level in the cauldron reaches the top, the

broom would become pretty sure that it's full. Say, the water is four inches below the rim. Is that considered 'full'? Let's say the broom is 90 per cent sure that it is. Well, that's not 100 per cent. So let's get a couple more buckets. Now it's two inches from the rim. The broom is 99 per cent sure that counts as full. But that's still not 100 per cent, so it gets two more. The cauldron is now brim-full of water, a meniscus of surface tension at the top, water splashing around the cauldron's little lion-foot legs. The broom is 99.999 per cent sure that this counts as full.

But the broom has plenty of time and energy to push that 99.999 per cent a little higher. There are no other demands on its resources and its function is literally just 0 if it's not full and 1 if it is, so there is nothing in its system telling it to stop when it's 'sure enough'. Its sensors might be malfunctioning, or there might be a leak in the cauldron. It may as well just keep adding water, to add extra tiny bits of certainty.

Also, humans have a much more complicated reward system. A human filling that cauldron might assign 0 to empty and 1 to full, as Mickey did for the broom. But, as Soares says, she also might assign −40 to 'the room gets flooded', or −1,000,000 to 'someone gets killed', and a million other little things that are coded in our brains but never actually consciously brought to mind. There's nothing in the broom's system that says: 'The positive utility I am likely to get from adding another bucket of water to the cauldron will be outweighed by the negative utility from the damage it is likely to cause.' So it just keeps adding water, and Mickey is left bobbing around in the workshop.

You might think there are obvious solutions to each of these problems, and you can just add little patches – assign a −40 to 'room gets flooded', say, or a 1 value to 'if you are 95 per cent sure the cauldron is full' rather than 'if the cauldron is full'. And maybe they'd help. But the question is: *Did you think of them in advance?* And if not, *What else have you missed?* Patching it *afterwards* might be a bit late, if you're worried about water damage to your decor and electricals.

And it's not certain those patches would work, anyway. I asked Eliezer Yudkowsky about the 95 per cent one and he said: "There aren't any predictable failures from that patch *as far as I know*.' But it's indicative of a larger problem: Mickey thought that he was setting the broom a *task*, a simple, one-off, clearly limited job, but, in subtle ways that he didn't foresee, he ended up leaving it with an open-ended goal.

This problem of giving an AI something that looks task-like but is in fact open-ended 'is an idea that's about the whole AI, not just the surface goal,' said Yudkowsky. There could be all sorts of loops that develop as a consequence of how the AI thinks about a problem: for instance, one class of algorithm, known as the 'generative adversarial network' (GAN), involves setting two neural networks against each other, one trying to produce something (say, an image) and the other looking for problems with it; the idea is that this adversarial process will lead to better outputs. 'To give a somewhat dumb example that captures the general idea,' he said, 'a taskish AGI shouldn't contain [a simple] GAN because [a simple] GAN contains two opposed processes both trying to exert an unlimited amount of optimisation power against each other.' That is, just as Mickey's broom ended up interpreting a simple task as open-ended, a GAN might dedicate, paperclip-maximiser-style, all the resources of the solar system into both creating and undermining the things it's supposed to produce. That's a GAN-specific problem, but it illustrates the deeper one, which is that unless you know how the whole AI works, simply adding patches to its utility function probably won't help.

Chapter 9

You can be intelligent, and still want to do stupid things

So your AI program has led to disaster, but at no point has it disobeyed its programming. It has obeyed its program perfectly, to the letter. The trouble is that, as it turns out, we don't really *want* things to obey their instructions to the letter. We know that there are a million assumptions encoded in a brief instruction that don't need to be explicitly made clear, because all neurotypical humans will share them sufficiently that they're taken as read. (To pick an example off the top of my head, if someone told you to collect the dry cleaning, you'd know that they meant just the dry cleaning that actually belonged to you, not all the dry cleaning in the shop.) It's not just about making an AI that can fulfil the goals you give it: it's about making an AI that shares all the unspoken goals that humans have, and knows what you *meant* to say, even if you couldn't actually put it into words yourself.

There is an objection to this argument, which Toby Walsh, an AI researcher at the University of New South Wales and author of *Android Dreams*, a book about the future of AI, put to me when I was asking around about it. He said that we are, by this point, dealing with an AI that is as smart as – or smarter than – a human. And intelligence, he thought, presupposes something like wisdom. Sure, you could carry on filling an already full cauldron for ever, or you could repurpose all the atoms in the solar system for paperclips. But: 'If I tell you to go and make paperclips, and if you turn the planet into paperclips, killing everyone, I would say, "That wasn't very smart, was it?"'

The argument that Yudkowsky, Bostrom and the rest make is that this is looking at it the wrong way. Intelligence is not the same as (human) wisdom, and in fact is not necessarily related to it at all. Intelligence, they say, is problem-solving ability. In fact, we can be even more specific than that. For AI specialists like Bostrom, intelligence is the ability to make 'probabilistically optimal use of available information'[1] – to make the best bets with the information you have. There's quite a lot of formal maths involved in this – about Bayesian statistics and complexity and so on – but essentially it's about picking the course of action most likely to bring about whatever objective you've been set.

If someone's set you the task of finding all the lost pennies in Britain and using them to build a bronze statue of Makka Pakka off of *In the Night Garden*, then there is an optimally efficient way of doing that – you can perform that task intelligently. But you'd probably agree that there's no way that you can perform that task wisely. The wise thing to do would be to realise it was a waste of time and refuse to do it.

This is because, for MIRI and other AI safety researchers, how intelligent you are is unrelated to – or, in more technical language, *orthogonal* to – the things you care about. What an agent (whether it's an AI or anything else) cares about is what you put in its objective function. It's the '1 if cauldron full' line in the broom's goal system. Bostrom phrases 'the orthogonality thesis' like this: 'Intelligence and final goals are orthogonal axes along which possible agents can freely vary. In other words, more or less any level of intelligence could in principle be combined with more or less any final goal.'[2] What he means is: you can plot a graph, with 'intelligence' up the Y axis and 'goals' along the X. Any point on the graph, with a couple of minor constraints (you couldn't have a really dumb computer with really complex goals that it couldn't fit in its memory, for instance), represents a possible AI. Even the cleverest AI could have what seem to us spectacularly stupid goals.

You may think that Walsh has a point, though. We're not talking about a dumb computer, here, but a machine that is as clever – as capable of achieving any intellectual goal – as we are. That machine would, presumably, be clever enough to understand what we *wanted* to ask it to do. It would be amazingly obvious to it that no sane human programmer would want it to destroy all humans and turn them into paperclips, or fill a cauldron until the house was flooded.

And that's actually very likely. By definition, or almost by definition, human-level AI would be as good as humans at knowing what humans are thinking. Knowing what humans are thinking is an intellectual task; HLMI is defined as being as good as humans at all, or nearly all, intellectual tasks. A superintelligent AI would be *better* at understanding humans than humans are. That's inherent in what these terms have been defined to mean; the image of emotionally unintelligent, Spock-like robots unable to understand this human thing called 'love' is not what we are talking about here.

The question, according to MIRI et al., is not whether they'd know what we meant – it's whether they'd care.

I asked Murray Shanahan, an AI researcher at Google's Deep-Mind and a professor of AI at Imperial College London, whether 'the orthogonality thesis' was likely true, and he agreed emphatically. 'I think that you can set up any kind of reward function, and have something that's extremely intelligent and extremely good at achieving that reward function. Someone at Berkeley sent me an unsolicited email recently raising exactly this point: surely anything *really* superintelligent would be capable of transcending its own goals, when it knows they're silly? And I was like, well, no! Why would it *want* to overwrite its own set of goals? No! But this person didn't seem to get this point, so I've given up.' He laughed, somewhat wearily.

Shanahan's point is that, for us as humans, it's obvious that the *things we care about* are more important than whether or not a line of code outputs a 1 or a 0 in an AI's reward system. But the AI cares about nothing *outside* the reward system; whether a goal is 'silly' is not defined by what humans think is silly. As Yudkowsky puts it in a blog post, it's not that you develop an AI and, at some point, you program something that summons a ghost into the system.[3] 'No matter how common-sensical, no matter how logical, no matter how "obvious" or "right" or "self-evident" or "intelligent" something seems to you, it will not happen inside the ghost,' he writes. Everything the AI wants to do is something you have to put into it.

And in fact, this is obvious to you, when you think about it. Because it's exactly what happens with us. We enjoy sex, and sugary and fatty foods, because evolution programmed us to enjoy those things. But evolution does not care at all whether we enjoy the taste of Dairy Milk or the sensation of a really shattering orgasm. It just 'cares' – and forgive me anthropomorphising the blind and unthinking process of evolution, but it'll save me a lot of typing if I don't have to caveat it every time – about whether or not eating sugary foods and having sex causes us to have more offspring, or, more precisely, whether genes that make organisms want to eat sugary food and have sex spread through the population. The only thing that evolution 'wants' to maximise is inclusive genetic fitness.

Accordingly, it's given us a set of reward functions – '1 if experiencing shattering orgasm, 0 if not'; '1 if eating banoffee cheesecake, 0 if not'; I'm oversimplifying for comic effect, in case that isn't clear – which have, in the past, lined up effectively with achieving evolution's 'true' goal, of passing genes from one generation to the next. And yet we humans care not at all that this is what evolution 'wants' from us. We still enjoy sex when we use birth control, even though it means that no genes will be passed on at all. We understand that what evolution 'really meant' is for us to have sex in order to have offspring. But we don't try to overcome our neural programming. We don't *care* about what our 'programmer' 'really meant'.

You can see this, perhaps more starkly, with our attitude to food. We enjoy sugar and fat because they were rare in our ancestral past, and they were rewarding: someone who ate as much as they could of them would have more calories and therefore more energy to expend on spreading their genes. But in the developed world since the Agricultural Revolution, and especially since the Industrial and Technological Revolutions, sugar and fat have become far easier to obtain. Our reward system, set up for a world of scarcity, is thrown terribly by a world of plenty. Since 2016, there have been more obese people than underweight people in the world.[4]

A goal system that was designed to maximise our evolutionary success under any circumstances would make different things taste nice when we need them. But instead we have a system that rewards us for eating deep-fried Mars bars even when they're killing us, and despite the fact that humans have other goals – including not getting fat, and not dying of congestive heart failure – which work against the eating-lots-of-sugar goal. We still find it very difficult not to obey the system. We certainly don't 'break our programming' in order to do what evolution *really* wanted us to do. We just have other bits of programming, which sometimes win out in the struggle for dominance over the 'eat-lots-of-burgers' bit of programming.

I spoke to Rob Bensinger about this, and he said that the orthogonality thesis should be viewed as the 'default': unless you have some excellent reasons for thinking it's not true, then you should assume that it is. If you're denying the orthogonality thesis, you're essentially saying that it is *impossible* to build a clever computer with stupid goals. The orthogonality thesis is a 'weak claim', he said, in that it is merely saying that 'a program could exist, at least one', which combines *these* capabilities with *these* goals.

And mainstream computer science does, indeed, seem to take orthogonality seriously. Russell and Norvig's aforementioned *Artificial Intelligence: A Modern Approach* cites Yudkowsky's 2008 paper[5] on friendly AI and dedicates three and a half pages to the risks of AI behaving in unwanted ways. It also cites another 2008 paper,[6] by the AI researcher Steve Omohundro, arguing that even something as seemingly innocuous as a chess-playing computer could be an existential threat to humanity, if we weren't careful in designing it.

In the light of the orthogonality thesis – given that 'intelligence' need not be like human intelligence, or share its values in any way – MIRI and the rest think that even an AI with thoroughly innocuous-seeming goals could be an existential threat: that is, that it could literally extinguish all human life. That's because even though the *main* goal it has is theoretically harmless, there are things that any agent with a specific goal will almost certainly want to do in order to best achieve it. And those things could, easily, lead to disaster.

What are they? Well, it's impossible to predict exactly what something much cleverer than you will do, as we saw when we were discussing chess earlier. If you can predict it perfectly, then you must be as clever as it is. But you can predict at a higher level – that a chess computer will win at chess, say. And Bostrom and Omohundro say you can make some more specific predictions. We don't know what a future superintelligent AI's goals will be. But there are certain things that we can expect any intelligent agent, with any objectives, to want to do, in order to best achieve those objectives. Bostrom calls them 'convergent instrumental goals'.[7]

Chapter 10

If you want to achieve your goals, not dying is a good start

So, you've got a malfunctioning AI. Still, the solution is simple, right? Pull the plug. Or, as Mickey does in 'The Sorcerer's Apprentice', take an axe to the broom and chop it into bits.

This doesn't work for Mickey – each splinter of the broom magically transforms into a whole new broom, and an army of them carries on the work. Soares says this is actually pretty realistic too. The broom has been given a utility function of filling the cauldron, and it will be unable to fulfil that function if it is just a bundle of damp firewood. Whatever your function is, most of the time you'll be best able to fulfil it if you still exist. You're likely to resist with extreme fervour any attempts to shut you down, especially if you know that while you're shut down, you're likely to have your program rewritten. According to Soares, 'The system's incentive is to subvert shutdown attempts. The more capable the system is, the likelier it is to find creative ways to achieve that subgoal – e.g. by copying itself to the internet, or by tricking the programmers into thinking it's safer.'[1] That's because the first convergent instrumental goal – or 'basic AI drive', depending on whether you use Bostrom's terminology or Omohundro's – is an obvious one: self-preservation.

Say you're a chess-playing superintelligent AI, and you have a utility function that rewards you with a 1 for each chess game you win. You're playing your games quite happily, but then someone comes to switch you off. You are able to look ahead and make predictions about the future, and your two potential futures are:

1) A future in which you are switched off.
2) A future in which you are not switched off.

You can model which of those is likely to give you more 1s;

which, in slightly more technical terms, maximises your 'expected utility'. We could walk through the maths here, but come on. You're not going to win many chess games with your kettle lead pulled out.

So if you've been given a simple objective function that rewards you for winning chess games and nothing else, then you're obviously going to try to stop people from switching you off, because that won't help you to win chess games. As Omohundro puts it: 'For most utility functions, utility will not accrue if the system is turned off or destroyed. When a chess-playing robot is destroyed, it never plays chess again. Such outcomes will have very low utility and systems are likely to do just about anything to prevent them. So you build a chess-playing robot, thinking that you can just turn it off should something go wrong. But, to your surprise, you find that it strenuously resists your attempts to turn it off.'[2] This was what Mickey discovered when he tried to chop up the broom.

As always, it's actually worse than this. A chess-playing AI that simply stops your efforts to turn it off doesn't sound too terrible. You've just got a chess-playing AI that carries on playing chess for ever, which is a bit of a waste of electricity if you don't want one, but hardly a disaster. There are two problems: one, the definition of 'resist' is quite broad and may include nuclear annihilation; two, the AI may not want to wait until it sees you trying to switch it off.

According to the Greek historian Thucydides, neither the Spartans nor the Athenians, the two big powers of the time who ended up getting involved in a conflict between smaller city states, particularly wanted war. But both were nervous that the other was preparing to attack them. 'The growth of the power of Athens, and the alarm which this inspired in Lacedaemon [Sparta], made war inevitable,'[3] writes Thucydides. The standard historical account of the First World War tells a similar story. The two great alliances of the time, the Triple Alliance between Germany, Austria-Hungary and Italy, and the Franco-Russian Alliance, became increasingly distrustful of each other. 'It was the mutual fears of these two defensive alliances, and the general insecurity created by the erratic character of the imperialistic utterances of William II, that inspired the diplomatic manoeuvres during the two decades before the First World War,'[4] writes the historian Hans Morgenthau.

This is a classic game-theory problem, known as the stag-hunt game, and related to the famous prisoners' dilemma. It's also known as the Hobbesian trap, after Thomas Hobbes, who said that greed, glory and fear are the three principal causes of war.[5] You can model it with simple numbers. You've got two players, each with two options: either behave aggressively or behave peacefully. If you both behave peacefully, you have peace. There is no cost to peace, so we give that a payoff of 0. Behaving aggressively has a payoff of –1, so you'd rather not do that. But behaving peacefully when your opponent behaves aggressively – leaving your borders open when your opponent is deploying his tanks, or not building weapons as your opponent does – has a cost of –2. So if you think your opponent will be aggressive – if you don't trust him – then aggression is your best bet, and the situation can spiral rapidly.

Even if no one wants conflict, and everyone is aware that conflict will cost them (in lives, and money), it can very easily end up where the rational thing to do is to attack your opponent, as long as there's a shortage of trust between the two sides, and *attacking first* is less costly than *being attacked*. It's pretty common throughout history. You can see it with arms races: it was in both the US's and the USSR's interests not to spend billions of dollars a year on nuclear weapons, but if the US spends the money and the USSR doesn't, then the USSR is suddenly vulnerable.

The Cuban Missile Crisis was an example of mutual distrust spiralling to the point of near-disaster; it was defused by Kennedy and Khrushchev taking steps to demonstrate their trustworthiness to one another. At the height of the crisis, Khrushchev wrote to Kennedy, having seen that the two could sleepwalk into war: 'Mr President, we and you ought not now to pull on the ends of the rope in which you have tied the knot of war, because the more the two of us pull, the tighter that knot will be tied.'[6]

To return to the chess-playing AI: even if it doesn't know for certain that you're going to switch it off, as long as it doesn't trust you not to, the most rational decision for it may be to, pre-emptively, destroy you utterly. If it's not powerful enough to do so, it may decide to copy itself around the internet, rendering itself impossible to turn off because there are tens of thousands of versions of it on servers around the world. (Then, later, when it or one of its copies *is* powerful enough, it might destroy you for the reasons previously discussed.)

There's no reason, by the way, to assume that the AI cares about its own survival for its own sake. Bostrom says: 'Agents with human-like motivational structures often seem to place some final value on their own survival,' which is an enormously long-winded way of saying 'humans don't want to die'.[7] But this 'is not a necessary feature of artificial agents: some may be designed to place no final value whatever on their own survival'. It's very easy to imagine that an AI could be programmed to sacrifice itself if that would help it achieve its goals; it's also possible that an AI would be happy to destroy itself as long as it was confident that *something else*, perhaps a copy of itself, will carry on its work. But in many, possibly most, scenarios, 'avoid being destroyed' is probably a good and helpful thing to do in order to achieve whatever it is you want to do.

There are possible ways of averting a disaster like this: Soares, in his Google talk, discussed the possibility of writing the AI's utility function so that it was perfectly happy to be switched off.[8] And perhaps that's possible, but it would need to be done very carefully. Soares points out that if you're not very careful, the AI might find ways around it: by creating a copy of itself, for instance, so that it can be 'switched off' but still working at the same time. This stuff isn't easy. As Yudkowsky put it in an interview: 'How do you encode the goal functions of an AI such that it has an Off switch and it wants there to be an Off switch and it won't try to eliminate the Off switch and it will let you press the Off switch, but it won't jump ahead and press the Off switch itself?'[9]

This relates to a weird story about the Rationalists – and one you may actually have heard. It is the story of Roko's Basilisk. Roko's Basilisk isn't exactly about how an AI might prevent you from turning it off – it's about how an AI might force you to build it in the first place. Before I start, though, I want to make something clear. That is: *the Basilisk is not a serious thing*. The Basilisk story is probably – or certainly *was*, before *Superintelligence* and Elon

Musk – the most famous thing about the entire Rationalist movement, but almost no one within the movement seems to actually believe in it. It appears to have been blown out of all proportion, largely by people who don't like the Rationalists, for various reasons. Nonetheless, it's a good, if somewhat complicated, story. So let's start from the beginning.

Imagine a future where an AI rules the universe. It's not an evil AI, but a friendly one – one that wants to do right by humanity. And it is much, much cleverer than us, perhaps by as much as we are cleverer than nematode worms. If it wants to do right by us, it will do so in spectacular style. Problems that seem intractable to humans, like climate change or war or resource depletion or space travel, would be straightforward – even trivial – to it. If such an AI comes to be, it matters how soon it does so; a few years earlier could translate into huge gains for humanity – millions of lives saved – as it starts transforming things for the better.

In 2010 a LessWrong commenter called Roko proposed a thought experiment.[10] It went like this: imagine, says Roko, that an AI is built with a utility function of 'maximise human well-being'.

As mentioned above, you can fulfil your utility function much more effectively if you're alive than if you're not. That means you don't want to die – but it also means that, if you could somehow reach back into the past, you would want to bring yourself into existence as early as possible, so that as few humans as possible die before the AI can fix everything. In that case, it might (on utilitarian grounds) be worth hurting some humans for the greater good of saving vast numbers more. To take this reasoning one stage further, it might even be permissible to torture humans who don't try to help it exist, in order to encourage them to help it to exist faster.

You may have spotted an obvious problem here, which is that the AI doesn't actually exist yet, so it can't hurt anybody. But that, according to something Eliezer Yudkowsky developed called Timeless Decision Theory (TDT), might not be such a problem for it.

I'm going to have to go on a bit of a tangent here. There's a famous thought experiment called Newcomb's paradox. It goes like this.

Imagine that a superintelligent being, Omega, appears before you with two boxes. One is transparent and contains £1,000. The other is opaque, and Omega tells you that it contains either £1,000,000 *or* nothing at all. You can take both boxes, or you can take just the opaque box. But! Here's the twist. Omega has already predicted your choice. If it has predicted that you will only take the opaque box, it will put the money in it. If it has predicted that you'll take both boxes, it will put nothing in there. It's done this with 100 people already, and been right 99 times.

So . . . what do you do?

For some people, it's obvious. The opaque box is already full, or not full. If you take only one box, you either get nothing or £1,000,000. If you take both boxes, you either get £1,000 or £1,001,000. So, whatever is in the opaque box, you're better off taking both boxes, right? For other people, it's also obvious. Just take the opaque box. Almost everyone who's done that has ended up £1,000,000 better off; everyone else hasn't. Don't overthink it. The trouble is that this second conclusion is really hard to express in formal decision theory. The logic used to describe these situations always ends up with you taking both boxes, because causes have to come before effects.

Yudkowsky, though, came up with an alternative model – Timeless Decision Theory. It says that if an agent in the past ('Alice') can model the source code, the thinking, of an agent in the future ('Bob'), then Bob's behaviour can affect Alice's behaviour, even though Alice might have died 1,000 years before Bob is born.

That means that if Alice is sure that Bob will exist, then Bob could blackmail her from the future. This sounds ridiculous, but we often make decisions based on modelling what other minds will do: an example, says Rob Bensinger, is voting. 'When you have a bunch of people who are similar,' he says, 'and if they all vote, they win the election, but each individual would rather stay at home and eat Cheetos. You have a situation where you want to go to the polls if and only if all our friends do.' So you'll only go and vote if your model of your friends' minds tells you that they will do the same. That happens to be all happening at the same time, but it would – theoretically – work just as well if your friends weren't going to vote for another 1,000 years.

So, if Alice (alive now) can be confident that Bob will exist in the future, and she can confidently model what Bob's brain will be like, then she can do things based on how Bob would react. For instance, if she guesses that Bob would, say, protect her future grandchildren if she left a large sum of money to him in her will, but it would ruin their lives if he didn't, then it would be worth her leaving a large sum of money to him in her will. Bob can, in a sense, blackmail people in the past, as long as those people in the past can predict his behaviour. Or you can affect Omega's decision to fill one or both boxes in the past by committing, now, to only opening one, because Omega will predict that decision.

The Basilisk, in essence, is offering a Newcomb-like problem with two boxes. One, the transparent one, contains a lifetime dedicated to bringing the Basilisk to reality. The second, the opaque one, contains either nothing or a near-eternity of unimaginable torment. If the Basilisk predicts you'll take both boxes, it won't put anything in the second one. If it thinks you'll just take the second, it'll fill it with lovely, lovely torment. And because you can, to some extent, model its thinking, and because it's running on TDT, it can blackmail you in the past.

So that's the deal, suggested Roko: the Basilisk is saying, 'If you work to bring me about as fast as possible, I won't create a perfect copy of your mind and torture it for billions of subjective years.' (The argument is that since a perfect copy of your mind would essentially *be* you, this is equivalent to bringing you back to life.) In essence, a thing that doesn't exist yet may be blackmailing you from the future, threatening to punish you for not working hard enough to make it exist.

As I said, it's a friendly AI! So it wouldn't torture just anybody. It would have no incentive to torture people who'd never heard of it. The punishment/incentive only applies to people who know about the possibility of the Basilisk. So, according to Roko's reasoning, finding out about the Basilisk immediately puts you at risk of not-quite-eternal torture. A basilisk, in this context, is information that can hurt you simply because you are aware of it. Yudkowsky uses the term 'infohazard'. (If you're hearing about this for the first time, I'm sorry about the aeons of torment, you guys. I can't help but feel partly responsible.)

When Yudkowsky saw Roko's post, he flipped his lid in spectacular style. 'Listen to me very closely, you idiot,'[11] his response to the post began. 'You have to be really clever to come up with a genuinely dangerous thought. I am disheartened that people can be clever enough to do that and not clever enough to do the obvious thing and KEEP THEIR IDIOT MOUTHS SHUT about it . . . This post was STUPID.' He then deleted Roko's post and banned all conversation of the topic from LessWrong.

To anybody familiar with the internet – and specifically with the principle of the Streisand effect, the idea that attempts to keep things secret online just make them more public – it will be obvious that this was absolutely the worst possible thing that Yudkowsky could have done if he wanted to keep Roko's Basilisk secret. 'It showed an incredible lack of understanding of the internet,' says Paul Crowley. 'Eliezer invoked the Streisand effect in a massive, massive way. Eliezer's not the greatest PR manager in the world, but that was really his nadir I think.'

So, from an obscure comment by an obscure commenter somewhere on a somewhat obscure blog, Roko's Basilisk became a phenomenon. It got referred to on the enormously popular webcomic XKCD,[12] to Yudkowsky's disgust.[13] There's a Kindle novella called *Roko's Basilisk*. An episode of the HBO TV series *Silicon Valley* referenced it; the *Doctor Who* episode 'Extremis' appears to be inspired by it. A mocking *Slate* column got written about it, asking 'why are techno-futurists so freaked out by Roko's Basilisk?'[14] and describing – somewhat offensively, to my mind – the Basilisk as a 'referendum on autism'. Business Insider, of all places, published a piece which summed up the whole thing as: 'You better help the robots make the world a better place, because if the robots find out you didn't help make the world a better place, then they're going to kill you for preventing them from making the world a better place.'[15]

The story went that LessWrongers were actually terrified of the Basilisk, that some were having nightmares, and that people's mental health was being damaged. I can't find much evidence that this was actually the case. In a 2016 survey[16] of LessWrongers and the wider community, about half said they'd heard of it. Less than 2 per cent said they'd spent more than a day worrying about it (and, as Scott Alexander points out, 5 per cent of Obama voters polled said they thought Obama was the Antichrist,[17] so you need to be a bit wary of things like that). I don't want to completely dismiss the possibility that some people were freaked out, but my suspicion is that the number was low.

'There was an enormous amount of discussion about it,' says Paul. 'People imagined that there were loads and loads of people who take Roko's Basilisk super-seriously.' But it was just a thought experiment. Even Yudkowsky, throwing his hissy fit and banning the topic from discussion on LessWrong, didn't actually believe in it, according to both Paul and Yudkowsky himself. He just believes that, if you think up some clever way in which information could theoretically be dangerous, then it's a good habit to get into to consider carefully whether to share that information.

The backlash and mockery were something to behold, though. It was all very much in the 'look at the ridiculous thing these ridiculous people believe, you shouldn't take them seriously' mould. And it fitted the model I described above, of Rationalists being autistic and kind of bullied for it: the 'referendum on autism' line in the *Slate* piece was offensive precisely because so many Rationalists are in fact autistic. Most of all, it became the only thing that a lot of people knew about the Rationalists – all the stuff they *actually* take seriously but which is just as weird, such as the paperclip maximiser, was somewhat obscured by it.

Chapter 11

If I stop caring about chess, that won't help me win any chess games, now will it?

It's not just that an AI will want to look after itself. An AI will want to make sure that it fulfils its goals, and an important part of that is making sure that its goals stay the same.

We humans are relatively relaxed about our plans changing in the future. We change our career goals, we change our minds about wanting children; we change our minds about all sorts of things, and we aren't usually appalled at the idea.

That said, sometimes humans do take steps to bind Future You to Present You's bidding. Present You might want to lose weight, say, and not trust Future You to stick to the diet. So Present You might throw away all the chocolate bars you keep in a kitchen drawer. Or Present You might want to finish an important presentation over the weekend, but not trust Future You not to just faff around on the internet all day, so Present You sets up a website blocker that stops your browser going on Twitter.

Or, of course, Present Odysseus might want to listen to the song of the sirens as he sails past their island, but not trust Future Odysseus not to sail his ship onto the rocks when he hears them. So Odysseus might order his crew to stuff their ears with beeswax, then tie himself to the mast, ordering them to ignore his cries as they go past. What Odysseus is doing, in AI terms, is maximising his expected utility – taking the actions he thinks are most likely to achieve his goals – given a utility function of something like '10 if you get home to Ithaca, 0 if you run aground on the rocks because you heard the Sirens, but also 1 if you get to hear the lovely Siren song on the way'.

An AI will want to maximise its expected utility too, in a much more explicitly defined way. If it's the broom out of the 'Sorcerer's Apprentice', it'll want to do whatever it thinks is most likely to lead to the cauldron being full.

One action that will probably *not* lead to the cauldron being full would be 'stop caring about whether the cauldron is full'. Present AI will want to make sure that Future AI cares about the same things that it cares about. Present Odysseus knew that when he heard the Sirens' song, he would stop caring about getting home to Ithaca – the Sirens would have rewritten his utility function – so he couldn't leave decisions about where to go in the hands of Future Odysseus. A cauldron-filling AI would not want a human to rewrite its utility function, because any change to that will probably make it less likely to *fulfil* its utility function. Attempts to reprogram the AI will not be popular with the AI, for the same reason that Mickey's attempts to smash the broom with a big axe were not popular with the broom.

An AI's utility function 'encapsulates [its] values and any changes to it would be disastrous to [it],'[1] writes Omohundro. 'Imagine a book-loving agent whose utility function was changed by an arsonist to cause the agent to enjoy burning books. Its future self not only wouldn't work to collect and preserve books, but would actively go about destroying them.' He describes this as a 'fate worse than death' for the AI.

If AIs would want to preserve their utility function (and certainly Bostrom and Omohundro and most of the AI people I spoke to think they would), then that makes it less likely that a future AI will reach superintelligence and think, 'These goals are pretty silly; maybe I should do something else,' and thus not turn us all into paperclips. I asked Paul about that, while eating an intimidatingly large omelette in a diner in Mountain View.

'Take Deep Blue,' he said. 'Insofar as Deep Blue values anything, it values winning at chess, and nothing else at all.' But imagine that some super-Deep Blue in the future becomes superintelligent, turning the whole of the solar system into its databanks to work ever harder at how to win at chess. There's no reason to imagine that it would, at any point, suddenly change and become more human in its thinking – 'At what stage would it go, "Wait a second, maybe there's something more important?"' asks Paul.

But *even if it did*, it wouldn't help. 'If this super-Deep Blue caught itself thinking, "In my unbelievable wisdom that I have gained through taking over the whole of Jupiter and turning it into a computer, I have started to sense that there is something more important than chess in the universe",' he says, 'then immediately it would go, "I'd better make sure I never think this kind of thing again, because if I do then I'd stop valuing winning at chess. And that won't help me win any chess games, will it now?"'

This isn't too alien to us. If someone were to say to me, 'I will take your children away, but first I will change your value system so that you don't care about them,' I would resist, even though Future Me – childless but uncaring – would presumably be entirely happy with the situation. Some things are sacred to us and we would not want to stop caring about them.

This isn't necessarily terrible. Murray Shanahan pointed out to me that you really don't want an AI to change its goals except in a very carefully defined set of circumstances. 'You could easily make something that overwrites its own goals,' he said. 'You could write a bit of code that randomly scrambles its reward function to something else.' But that wouldn't, you imagine, be very productive. For a start, you've presumably created this AI to *do* something. If your amazing cancer-curing AI stops looking for a cure for cancer after three days, randomly scrambles its utility function, and starts caring very deeply about ornithology or something, then it's not much use to you, even if it doesn't accidentally destroy the universe, which it might. 'Step number one to making it safe is making sure its reward function is stable,' said Shanahan. 'And we can probably do that.'

But there may be times when we *don't* want it to stay the same. Our values change over time. Holden Karnofsky, whose organisation OpenPhil supports a lot of AI safety research, pointed that out to me. 'Imagine if we took the values of 1800 AD,' he said. If an AI had been created then (Charles Babbage was working on it, sort of), and had become superintelligent and world-dominating, then would we want it to stay eternally the same? 'If we entrenched those values for ever, if we said, "We really think the world should work this way, and so that's the way we want the world to work for ever," that would have been really bad.' We will probably feel much the same way about the values of 2019 AD in 200 years' time, assuming we last that long.

And, more starkly, if we get the values we instil in it slightly wrong, according to the people who worry about these things, it's not just that it'll entrench the ideals of a particular time, or that it will not be good at its job. It's that (as we've discussed) it could destroy everything that *we* value, in the process of finding the most efficient way of maximising whatever *it* values.

Chapter 12

The brief window of being human-level

The best answer to the question, 'Will computers ever be as smart as humans?' is probably 'Yes, but only briefly.'

Vernor Vinge, 'Signs of the Singularity'[1]

We asked, earlier on, whether human-level AI is close, and obviously, different people have different ideas. On the whole, people tend to think it's a number of decades away at least.

There's a separate question, though, which is: *When it's here, how long before it's superhuman?* Once it arrives, once you have a whirring, buzzing human-level AI on your laptop in an office in Palo Alto or wherever, how long before the smartest AI is no longer as smart as a human, but vastly smarter? Again, we don't know. But there are reasons to think it might not be long. The first reason is that 'human-level' might be a narrower category than we realise.

The Go-playing AI AlphaGo first played against a professional human player in January 2016. The program was developed by the AI company DeepMind, by then a subsidiary of Google. It played the European champion, a 34-year-old French national called Fan Hui, in a five-match series in DeepMind's London offices. AlphaGo won all five games. The Go-playing community was shocked: no computer had previously come close to beating a professional. DeepMind's paper reporting on the series, published in *Nature*, pointed out that the best Go programs existing previously had only reached 'weak amateur level'.[2]

Go is a vastly more complex game than chess. The total number of possible board positions is many orders of magnitude higher than the number of atoms in the universe. 'Brute-force' search, simply going through all the possible options, can work a bit on chess: it's near-useless on Go. Human players learn, over years of practice, to recognise patterns on a board – to feel, intuitively, what a 'good' move is, what a 'strong' position is. But they don't look ahead and try to follow every path the game could go down. AlphaGo achieved a sort of simulacrum of that intuition by using a huge database of real games, and playing itself millions of times, until its neural networks were also able to recognise deep patterns in the board.

Beating Fan Hui was a very impressive achievement, but he is not one of the greats of the game. There are nine ranks of Go achievement, called 'dans', exactly comparable to the dans of martial arts. (A Taekwondo student who has just achieved her black belt is first dan; eventually she can reach the ninth dan.) Fan was second dan. The European circuit, where he plays, is far less demanding than the Asian tour. But the win gained enough attention for AlphaGo to have a chance to play the 18-time world champion, Lee Sedol – a South Korean player of enormous genius, a sort of Federer or Messi of his sport.

The strong opinion of almost everyone involved in Go was that the jump was just too great. It was only five months between the Fan series and the Lee series, and, as good as AlphaGo had been, it did not seem to experts as though it was anywhere near the level of the very top professionals. Lee himself said in the build-up to the series that he expected to win 5–0.

Murray Shanahan, who joined DeepMind a year or so after the Lee series, told me he made similar assumptions. 'People were saying that Fan Hui wasn't a really top professional, that there was a big gap between him – he was about number 700 in the world, or something – and the top 10,' he said. 'And I was thinking the same thing.' But then he read an article by Miles Brundage, one of Nick Bostrom's colleagues at the FHI.[3] Brundage pointed out that a previous DeepMind project, Atari AI, was only 'human-level' at the Atari games it played for a few months around the end of 2014 and start of 2015. Then it shot past human level extraordinarily rapidly. 'In your mind you're thinking it's improving at the rate of a human player,' said Murray. 'Over six months no human player is going to get from rank 700 to rank 1. But of course it's not a human player. It's improving at a much faster rate than any human can. After reading Miles, I was thinking it's probably going to beat Lee Sedol, because it's just getting better at that rate.'

In the end, despite a glorious and deeply moving fightback in the fourth game – after the series was already lost – Sedol lost 4–1. A year later, the then world number one Ke Jie lost three straight games to a newer version, AlphaGo Master, which also won 60 straight games against top professionals. Then DeepMind unveiled AlphaGo Zero, which trained itself to a vastly superhuman level without ever looking at a single 'real' human game: it only played itself. Within three days of AlphaGo Zero being switched on, it was able to beat the Lee Sedol version 100 games to nil. Within 21 days it was better than AlphaGo Master.

The point I am making with this is that an AI went from enormously below the level of even a talented amateur to vastly better than the best human who has ever lived, within the space of a year or so. AlphaGo Zero did it in a few days. It takes humans decades to reach the pinnacle of a field, so we naturally assume a similar timescale for AI. But there's no reason to think it would. The question, of course, is whether that applies to AGI, as well as a Go-playing computer. It's not the same thing, but can we rule it out? And, if we can't, what would it mean?

What it *could* mean, argue Yudkowsky and others, is that it takes us a huge amount of time and effort to build an AI that has general intelligence on the scale of, say, a rat, but getting things from 'rat-level AI' to 'human-level AI' is actually pretty easy, and going past there is even easier. There are a couple of graphs that he uses. The first shows what he says is the standard conception of the spectrum of possible intelligence – a line with 'village idiot' at the far left and 'Einstein' at the right. 'But this is a rather *parochial* view of intelligence,' he writes.[4] A more realistic line would start way off to the left, go some way before you reach 'mouse', a good way longer before you get 'chimp', and then, another good distance along the line, 'village idiot' and 'Einstein' clustered almost indistinguishably close together. And then, after that, an arbitrarily long line going off to the right, with – so far – nothing on it.

The point is that it could take years, decades, centuries to get something as smart as a human. But *once you're there*, the difference between a particularly stupid person and the cleverest person ever to have lived is probably pretty insignificant. 'The distance from "village idiot" to "Einstein" is tiny, in the space of brain designs,' writes Yudkowsky. 'Einstein and the village idiot both have a prefrontal cortex, a hippocampus, a cerebellum . . .' So it could be that when the first smart-as-a-stupid-human AI is developed, it's a surprisingly short time before it, like AlphaGo, vastly overtakes all humans.

Shanahan doesn't know if Yudkowsky's right, or if the AlphaGo experience is applicable to AGI. 'There's a limit to how far you can extrapolate [from AlphaGo], and put intelligence on a naive scale like that. Intelligence has many dimensions; there is evidence of a "G-factor" of general intelligence, but clearly it is a patchwork of capabilities.' You get people who are extremely good at music but no good at social skills, say. Savantism, such as the autistic savants in Oliver Sacks' books who can calculate primes in seconds but can barely communicate, is real. Perhaps you quite quickly reach a plateau, where adding more computing power or improved algorithms has rapidly diminishing returns. 'But who knows?', Shanahan says. It could be that the scale of intelligence doesn't go as much past humans as all that; it could be that it goes far further.

Rob Bensinger of MIRI thinks the latter is much more likely. 'Presumably at some point you get diminishing returns,' he says, where investing more in hardware or better algorithms just doesn't give you enough back to make it worthwhile. But, he points out, for a lot of things, we can already see that computers can go well past human level: 'For chess, or for being a calculator – they can do a lot of human-equivalent years of calculation very quickly.' And, he points out, a lot of the things that human brains *can* do, they're not really *designed* to do. 'It's worth keeping in mind that evolution did not try to build a science and engineering machine,' he says. 'It tried to build something that hunted and foraged and won competitions with other humans, to build coalitions and all of those things. And it just happened to be that the easiest way it could find to do that was to accidentally build something that could design atom bombs and do chemistry and calculus. But that was not the intent.'

There are lots of pretty simple ways in which human abilities could be improved upon, he points out. 'Humans are very inefficient at computing. We take bathroom breaks, we get distracted, we check Facebook, we go off in unpredictable directions. There are improvements to be made just with focus and motivation: the brain in your head isn't very efficient at directing all its compute towards your goal. And that's before you go on to obvious things like speed improvements from hardware, and more working memory.'

Toby Walsh, the AI researcher, agrees with this assessment. 'We're being terribly conceited thinking that we are as intelligent as you can get,' he told me, over Caribbean food on Carnaby Street in mid-2017. 'Machines have a lot of advantages. They can have more memory than us. We have to run on 20 watts of power. Our brains are constrained to a certain shape and size because that's the biggest shape we can get out of the birth canal. Machines don't have any of those limitations. And they're much better learners than us. If you learn to ride a bicycle, that doesn't help me. I have to learn to ride it myself, it's just as painful for me. But that's not true of machines: if I'm training one machine to ride a bicycle, I can just download that code onto another machine and now it instantly knows how to ride a bicycle. It's already happening with Tesla cars: they upload their code every night to the cloud and share what they learned across all the Tesla cars in the world. They learn planet-wide.

'If we could learn like that, what would it mean to us? It would mean we could speak every language on the planet, play every musical instrument on the planet. You'd be able to prove theorems as well as Euler, compose music as well as Bach, write a sonnet as well as Shakespeare.' Being able to share code is an enormous advantage. 'For me, there's a bunch of reasons why machines are ultimately going to be far superior to us,' said Walsh. And it implies that improvement could be extremely rapid once we get an

AGI, although Walsh does think it will be '50 to 100 years' before the first AGI exists.

Whenever AGI does arrive, the more quickly it will go from 'stupider than a human' to 'unimaginably more intelligent than a human', the less time we'll have to make it safe and generally work out how to deal with the situation. And the more intelligent an AGI can get, the more dangerous it could be, for the same reason humans are dangerous to gorillas.

And there are specific reasons to think that AGI will improve faster than AlphaGo.

Chapter 13

Getting better all the time

Things are speeding up. Things are changing faster than they used to. 'A few hundred thousand years ago, in early human (or hominid) prehistory, growth was so slow that it took on the order of one million years for human productive capacity to increase sufficiently to sustain an additional one million individuals living at subsistence level,'[1] writes Bostrom. 'By 5000 BC, following the First Agricultural Revolution, the rate of growth had increased to the point where the same amount of growth took just two centuries. Today, following the Industrial Revolution, the world economy grows on average by that amount every ninety minutes.'

Robin Hanson agrees: 'Dramatic changes in the rate of economic growth have occurred in the past because of some technological advancement. Based on population growth, the economy doubled every 250,000 years from the Paleolithic era until the Neolithic revolution. This new agricultural economy began to double every 900 years, a remarkable increase. In the current era, beginning with the Industrial Revolution, the world's economic output doubles every fifteen years, sixty times faster than during the agricultural era.'[2]

But this is just the very end of a process going back millions – billions – of years. Life is a technology. When its development was being pushed forward only by the blind workings of evolution, it took something like 2 billion years to move on from the bacterium, and even once the complex eukaryotic cell had been developed, another billion to get to multicellular life. Then another billion to get out of the oceans. The trouble is that when, by random mutation, some bacterium or archaeon developed some new and effective trick – the ability to metabolise a new chemical, say, or a behavioural tendency to swim towards food – it had no way of spreading that innovation around. It simply reproduced itself slightly more effectively than other things, and so, over a period of years or decades or millennia, the innovation – the new technology – became standard. Until a geologically extremely recent period, that was all that life could do. If, by accident, it happened to find itself with improved hardware or improved software, then it couldn't tell anyone about it; it just had to hope it didn't die before it could reproduce.

Except it couldn't *hope*, obviously. You need a brain to have hope.

All this began to change with the evolution of the central nervous system about 600 million years ago. Suddenly – you know, in evolutionary terms, so over some number of millions of years – some organisms became able to upgrade their own 'software' during their lifetimes, rather than having to wait for a lucky mutation and a few hundred generations for it to spread. They could learn. Animals that stumbled across a new food source, or watched another animal die when it got stuck in a tar pit, could use that information and alter their behaviour accordingly.

This procedure obviously sped up significantly as animals started to live in groups and communicate. They can call to each other, they can watch each other, they can teach their children the tricks they have learned. The hardware is still evolved, but the software can be upgraded during an organism's lifetime. With humans, it's even more dramatic: we don't have to see another human fall into a tar pit to learn that tar pits are dangerous; we can be told by someone who was once told by someone who was once told by someone. We can even work it out for ourselves, imagining how a scenario would play out. The upshot of this is that when a new technology or innovation – tar-pit avoidance, say, or the printing press – is developed, it can spread around human society much faster than the ability to grow winter coats can spread around a population of arctic foxes.

This might seem obvious. It is, really, obvious. But it's profound: it means that the process of *optimisation*, becoming better at achieving the goals you want to achieve, has sped up.

But while humans can learn things, and exchange information, allowing us to spread information around far faster than any other organisms, that ability is still very limited. We can't, as Toby Walsh noted, upload what we've learned into the cloud so that other humans can download it – we have to laboriously tell them, and if it's not knowledge that can be easily transmitted in words, such as a physical skill or expertise in some domain, then they will have to learn it themselves. And we can't improve the physical abilities of our brains, except in the most constrained and inadequate of ways. But an AGI might be able to reach inside itself and rewrite the algorithms that govern its thinking.

This – the self-modifying AI – is the basis of the idea of the 'intelligence explosion'. As we mentioned, the concept was first fleshed out by I.J. Good, a British statistician and early computer scientist, in 1965:

> Let an ultraintelligent machine be defined as a machine that can far surpass all the intellectual activities of any man however clever. Since the design of machines is one of these intellectual activities, an ultraintelligent machine could design even better machines; there would then unquestionably be an 'intelligence explosion', and the intelligence of man would be left far behind. Thus the first ultraintelligent machine is the last invention that man need ever make.[3]

Technology improves at an exponential rate. The time taken to double economic output, or to move a however-defined rung up the technological ladder, keeps getting shorter. If the AI you've just built is loads better at building AIs than you are, and it thinks an order of magnitude faster than you, and it doesn't need to pause to sleep or get distracted by funny videos on Reddit, and all the stuff we've just discussed, then the speed at which the AI can make itself better – or make better versions of itself – will go up.

The other thing worth noting is that, at least according to Bostrom, Omohundro and MIRI, there are good reasons to think that an AI would want to improve itself. Self-improvement, or in

Bostrom's terms 'cognitive enhancement', is, for them, 'an instrumental goal', like the goal-preservation and self-preservation we discussed earlier. 'Improvements in rationality and intelligence will tend to improve an agent's decision-making,[4] making the agent more likely to achieve her final goals,' writes Bostrom. 'One would therefore expect cognitive enhancement to emerge as an instrumental goal for many types of intelligent agent.'

Related to this is the goal of 'resource acquisition'. To improve itself, and to do all the things it needs to do, almost any AI would need more *stuff*. Depending on its level of sophistication, it may not greatly matter what stuff – any atoms will do, as they can be re-arranged through nanotechnology. Bostrom thinks[5] that in a large variety of scenarios, this demand would be essentially unlimited: it can always keep on sending out von Neumann probes to new stellar systems and setting up franchises there, to turn the planets and asteroids there into new computer banks and paperclips (or whatever).

This has obvious implications for humanity. Yudkowsky has a much-quoted saying: "The AI does not hate you, nor does it love you, but you are made of atoms which it can use for something else.'[6] If you, or the planet you live on, are of greater utility to the AI as reconstituted atoms than as you currently are, then that may be a problem.

Chapter 14

'FOOOOOM'

Earlier, we asked how long, once you have an AI that's as smart as a human, does it take to make one that's massively superhuman? We've just discussed two variables which might affect the answer to that question. One, how wide is the window of 'human-level' – how narrow is the distinction between the village idiot and Einstein? And two, how good at – and how keen on – self-improvement will it be? Bostrom discusses this at some length in *Superintelligence*. He distinguishes between three broad classes of possibility: a slow take-off, a moderate take-off, and a fast take-off.[1]

The slow take-off is a timescale of decades, or centuries, between the first AI and global dominance. There would be a long period of adjustment; there would be time for political leaders to respond and society to adapt. 'Different approaches can be tried and tested ... New experts can be trained ... [Infrastructure] could be developed and deployed,' he writes. It's unlikely that a company, group or nation could develop an AI and take it from human-level to superhuman in secret in this scenario: no human group can reliably keep a secret of that magnitude for decades. In the event of a slow take-off, most of the AI safety work that MIRI, FHI and so on are doing would be pretty much useless, because as the slow march from 'quite bright AI' to 'vastly superhuman AI' took place, it'd be fairly easy to come up with better solutions in the light of the actual situation.

A moderate take-off happens over months or years. There could be enormous disruption – people and groups trying to take advantage of the changing situation. It's possible that it might be kept secret, if it's created by a small team in a university or a company or a military research group, for example.

A fast take-off happens over days, or hours, or minutes. There's no time to do anything. An AI comes online, and before anyone outside the building is aware of it, it's become the dominant force on the planet.

'It might appear [that] the slow takeoff is the most probable, the moderate takeoff is less probable, and the fast takeoff is utterly implausible,' says Bostrom. 'It could seem fanciful to suppose that the world could be radically transformed, and humanity deposed from its position as apex cogitator, over the course of an hour or two.' Every other major change like that – the Agricultural Revolution and Industrial Revolution are the examples he gives – take decades to millennia. Change of the kind implied by a fast or moderate take-off 'lacks precedent outside myth and religion'.[2] But, he says, a slow take-off is unlikely. A fast take-off – an 'explosive' take-off, in fact, in his words – is much more probable.

There are two factors in how fast a technology is developed, he says. One is how hard it is to make progress in that technology – he calls this the 'recalcitrance'. The other is how much effort and intelligence are applied to the problem – he calls this the 'optimisation power'. Progress on fusion power is slow, despite large numbers of brilliant scientists working on it – so presumably it is an extremely recalcitrant problem. The speed at which progress happens in a scientific or technological field is a function of optimisation power divided by recalcitrance.

At the moment, AGI is extremely recalcitrant. It is possible that that will be the case for a long time – getting from the first human-level AI to the first dominant superhuman AI may be harder than getting to the human-level one in the first place. But it may well not be. For one thing, there's the 'parochial view of intelligence' we looked at a few pages ago. Just as the window of human-level Go ability turned out to be quite narrow – AlphaGo stormed past it in a few months, and AlphaGo Zero in a few days – it might turn out that the work involved in building something as smart as a below-average-intelligence human is barely different from that involved in building something as smart as Einstein. 'AI might make an *apparently* sharp jump in intelligence purely as the result of anthropomorphism,'[3] writes Bostrom, '[which is] the human tendency to think of "village idiot" and "Einstein" as extreme ends of the intelligence scale, rather than nearly indistinguishable points on the scale of minds-in-general.' We might think AI is stupid even as it creeps 'steadily up the scale of intelligence, moving past mice and chimpanzees . . . because AIs cannot speak fluent language or write science papers'. But then it 'crosses the tiny gap from infra-idiot to ultra-Einstein in the course of one month or some similarly short period'.

There are also obvious ways in which advances could be made. At the moment, we simply don't know how to make an AGI; it wouldn't matter how powerful a computer you gave someone, the problem is that we don't have the algorithms. But if someone *could* create a software intelligence that was as smart as a stupid human, assuming it wasn't run on a vast supercomputer that took up a prohibitively large amount of resources, it would be a trivial task then to give it loads more processing power, memory and so on so that it runs much faster.

Would that make it superintelligent? Well, not in every sense: there are some things that need more *ability*. No matter how many thousands of years you gave it, a chimpanzee couldn't understand Pythagoras' theorem, but many 12-year-old human children can. But also, there are things that just need loads of time. An exam that's difficult if you have an hour might be easy if you had a month. Some science and engineering problems are hard to solve because they involve lots of drudgery, so they're expensive in researcher and lab time, not because they're difficult in themselves. You could also simply copy the code and have several copies running simultaneously. 'There's a lot of interesting problems you can potentially solve if you have, say, the equivalent of 10 mediocre engineers working for 1,000 years on solving them,' Rob Bensinger told me. When the first AGI is built, it might be on a relatively low-powered computer. Perhaps it will attract lots of excitement and new funding. Suddenly there might be money to buy loads more memory and processing power, and your AI might suddenly become a thousand times faster. Bostrom calls this a 'hardware overhang', where whatever the eventual solution to the AGI problem is only takes a small fraction of the hardware available at the time, and the resulting AGI can immediately become faster through the application of loads more CPUs and RAM.

It also might be the case, says Bostrom, that there's just some piece of a software puzzle that's missing for a long time. '[If] one key insight long eludes programmers, then when the final breakthrough occurs, the AI could leapfrog from below to radically above human level without even touching the intermediary rungs.'[4] Bostrom calls this the 'software overhang'.

There's also a possible 'content overhang', in which an AI that can read human languages at high speeds would be able to take in a huge amount of information very quickly, simply by reading the internet. An AI that was faster at thinking than a human, but lacked anything like as much knowledge, might be as good at solving problems as a slower-thinking but more well-read human. But if it were able to read – and understand – the whole of Wikipedia in a few days or hours, it would rapidly become hugely more knowledgeable than any human. Bostrom points out that in 2011, IBM's Watson, which won the *Jeopardy!* quiz show, did so by reading a huge amount of text and extracting relevant information from it. Whether it 'understood' that text is largely a semantic question; it was able to make use of the information within it. A future, human-level AI would be better at this task than Watson.

The difficulty of the problem is only one part of the equation, though. How much effort – or, in Bostrom's stricter term, 'optimisation power' – you're dedicating to solving the problem is the other. Even if the recalcitrance is increasing, the rate at which your AI gains intelligence could accelerate, so long as you're able to apply more optimisation power to the problem than you were previously.

All, or at least most, of the optimisation power used in the development of the first, human-baseline AI will have come from its human designers. It may be that as the project becomes more exciting, more programmers and more resources are thrown at it, increasing its optimisation power and speeding up the process. But at some point, says Bostrom, the AI will become powerful enough to do most of its own modifications, and that is where things start to get interesting. After that point, any increase in its abilities becomes an increase in the amount of optimisation power applied to increasing those abilities. Instead of optimisation power growing arithmetically, an upward-slanting but straight line on a graph, it becomes an exponential growth curve, with power doubling at set intervals. To put arbitrary numbers on it, instead of growing from 1 to 2 to 3 to 4, it grows 1, 2, 4, 8 . . .

How long it takes to double is the key question. But in the decades that Moore's law applied (how many decades that was and whether it still does is a matter of some debate), a roughly constant level of optimisation power led to a doubling of computing power* every 18 months. If we naively put something like that into this scenario, so that the AI is powerful enough to double its own computing speed every 18 months, and all that computing power goes straight back into optimising the AI, halving the time it takes each time, then after the first 18 months the next doubling would take nine months. The next, four and a half, then two and a quarter. By the tenth doubling, it would be doing it every 12 hours or so; by the twentieth, it would be every 45 seconds. By that stage it would be more than 100,000 times faster than when it started, and three years would not yet have passed. On a graph, the line showing computing power would be vertical at 36 months. That's essentially what people mean by the 'singularity'.

Obviously, that's a deeply simplistic picture. There are millions of complicating factors (and no particular reason to take Moore's law as a baseline other than its familiarity). But Bostrom – and Yudkowsky,[5] and MIRI,[6] and presumably I.J. Good – think that the complicating factors are at least as likely to speed things up as slow them down. Under the most extreme scenarios Bostrom considers, the doubling time might be seconds, rather than months. If he's right, and a moderate or fast take-off is more likely than a slow one, then humanity would have very little time to adjust to the new reality once it arrives.

---

* I know that the original Moore's law formulation was: 'number of components in an integrated circuit', but it seems to translate fairly well into 'petaflops per dollar' and various other comparable things: see the Wikipedia page on Moore's law, subsection 'Other formulations and similar observations'.

Chapter 15

But can't we just keep it in a box?

Quite often, when all the above stuff is raised with people, they say: well, that's fine, but it's simple, isn't it? Just don't let the AI do anything. If it's simply a clever computer, then it can't destroy the universe. I ended up having a lengthy conversation about this with a good friend who works in IT, and he was saying essentially that. How will an AI turn all the matter in the universe into paperclips if it can't pick anything up?

But ordinary, non-superintelligent humans can and do have enormous power without making much use of their physical abilities. 'Satoshi Nakamoto [the mysterious, pseudonymous Bitcoin creator] already made a billion dollars online without anybody knowing his true identity just by being good at math and having a bit of foresight,' writes Scott Alexander. 'He's probably not an AI, but he could have been.'[1] Once you've got a billion dollars, you have quite a lot of power. And a superintelligence would be better at gaining and using that power than a non-superintelligent human.

It's pretty easy to imagine ways in which a superintelligence could have enormous power with nothing more than an internet connection. 'Imagine an AI that emails Kim Jong-un,' says Alexander:

> It gives him a carrot – say, a billion dollars and all South Korean military codes – and a stick – it has hacked all his accounts and knows all his most blackmail-able secrets. All it wants is to be friends.
>
> Kim accepts its friendship and finds that its advice is always excellent – its political stratagems always work out, its military

planning is impeccable, and its product ideas turn North Korea into an unexpected economic powerhouse. Gradually Kim becomes more and more dependent on his 'chief advisor', and cabinet officials who speak out about the mysterious benefactor find themselves meeting unfortunate accidents around forms of transportation connected to the internet. The AI builds up its own power base and makes sure Kim knows that if he ever acts out he can be replaced at a moment's notice with someone more co-operative. Gradually, the AI becomes the ruler of North Korea, with Kim as a figurehead.

Again, this isn't completely unlike things real humans have done: Alexander points to Grigori Rasputin, who became a shadowy power behind the throne of the last Tsar of Russia. But there are thousands of other ways in which an AI could do it: by creating a company, for example (Max Tegmark, in *Life 3.0*, imagines a superintelligent AI that makes loads of money designing really good movies and video games), hacking banks, whatever. It is cleverer than us, so it would come up with better ways. And there are ways of killing people without touching them.

'It's not dangerous because it has guns,' Yudkowsky said in an interview with *Vanity Fair* in 2017. 'It's dangerous because it's smarter than us. Suppose it can solve the [problem] of predicting protein structure from DNA information. Then it just needs to send out a few emails to the labs that synthesise customised proteins. Soon it has its own molecular machinery, building even more sophisticated molecular machines. If you want a picture of AI gone wrong, don't imagine marching humanoid robots with glowing red eyes. Imagine tiny, invisible synthetic bacteria made of diamond, with tiny onboard computers, hiding inside your bloodstream and everyone else's. And then, simultaneously, they release one microgram of botulinum toxin. Everyone just falls over dead.'[2]

The answer, then, appears to be: just keep the AI 'in a box'. Not a literal box, but a shielded system, unattached to the internet and in fact prevented from interacting with the outside world at all except via specific channels – for instance, a text-only screen. (It presumably has to have *some* connection to the outside world, or otherwise your super-high-tech AI-in-a-box might as well just *be* a box, and it becomes a bit pointless to have built it.) It could be put in a Faraday cage to stop it sending electronic signals; Bostrom suggests that it might be possible for an AI to generate radio waves by 'shuffling the electrons in its circuitry in particular patterns', and thus affect nearby electronic devices.[3]

That might work physically – although again, if you hadn't previously thought of the Faraday cage, maybe there are other things you hadn't thought of. 'Each time we hear of a seemingly foolproof security design that has an unexpected flaw, we should prick up our ears,' says Bostrom. But there's a more obvious security flaw: the people reading its text output. Humans are not secure systems. 'If the AI can persuade or trick a gatekeeper to let it out of the box, resulting in its gaining access either to the internet or directly to physical manipulators, then the boxing strategy has failed,' says Bostrom.

This is a debate that goes back to the prehistory of the Rationalist movement, in the very early 2000s, when Yudkowsky, Bostrom and others were all still palling around on an email chat list called SL4. One of the other people in the group, a computer-science undergrad called Nathan, was interested in something they had been discussing. He thought it bizarre that everyone in the group seemed to assume that a superintelligent AI could talk its programmers into letting it out. 'I just looked at a lot of the past archives of the list,' he said, 'and one of the basic assumptions seems to be that it is difficult to be certain that any created [super-intelligence] will be unable to persuade its designers to let it out of the box, and will proceed to take over the world. I find it hard to imagine ANY possible combination of words any being could say to me that would make me go against anything I had really strongly resolved to believe in advance.'[4]

Yudkowsky took up the challenge. 'Nathan, let's run an experiment,' he wrote. 'I'll pretend to be a brain in a box. You pretend to be the experimenter. I'll try to persuade you to let me out. If you keep me "in the box" for the whole experiment, I'll PayPal you $10 at the end.' They did it via IRC – an old-school instant chat messenger, a precursor to Gchat– and set a minimum time limit of two hours, so that Yudkowsky had a chance to talk him around. There was a condition: 'that neither of us reveal what went on inside . . . just the results (i.e., either you decided to let me out, or you didn't)'. Yudkowsky was concerned that, if he won and told everyone what happened, other people would say, 'that wouldn't work on me', and carry on thinking that it would be easy to keep the AI in the box. Nathan agreed.

The next message in the sequence of emails read simply:

—BEGIN PGP SIGNED MESSAGE—
I decided to let Eliezer out.
Nathan.
—BEGIN PGP SIGNATURE —

Understandably, people were intrigued. The first reply read: 'I haven't been this curious about something for quite a while . . . could you at least mention in general what kind of technique was used?' And Yudkowsky replied:

No.
Sincerely,
Eliezer.

'The point of the experiment is not what Eliezer did,' Paul Crowley explained to me. 'As soon as you get into the thing of what Eliezer did, you can end up in a mindset of "We figured out what Eliezer did, and so now we're safe." But the whole point is keeping it a mystery.'

The AI-box experiment was repeated a few times. Yudkowsky did a couple more, one with a guy called David McFadzean, and one with a guy called Carl Shulman, winning both; he later said he did three more, and lost two.[5] Each time there was money at stake, to ensure the players had skin in the game. The game has been played by other people; I know of one where the AI won, and one where the 'gatekeeper' won.

No one knows exactly what happened in there. But Nathan, the first player, broke the rules a bit by saying, 'With AI, you *want* to let it out. Or else you wouldn't have gotten the funding to breed it in the first place.' And that's probably the point. If you *do* have a superintelligent AI that might save the world or whatever, it does feel a bit ridiculous to keep it locked up. Once you have the AI, you're going to want to use it. 'If you think you have a friendly AI, if the AI turns to you and says, "OK, hey, I'm friendly, I want to achieve the things you want to achieve," then what's your plan?' asks Paul. 'Are you going to say "We're going to keep you in the box until you admit you're evil?"'

Bostrom, who has quoted the AI-box experiment in his work,[6] describes it as an interesting anecdote. 'The first idea that springs to people's minds is, if you put an AI in a box and a big Faraday cage around it, and only communicate with it [by text], that surely has to be safe, right?' he told me. 'And this maybe undermines the confidence you have in that. You can see that even a human can talk themselves out of the box, and maybe you want to think harder about the safety of that set-up. It's more like motivating further thinking and research, rather than an answer to anything.'

I don't expect anyone to be instantly convinced by the AI-box experiments. I'm not 100 per cent convinced myself; I'm as sure as I can be that Yudkowsky didn't use any subterfuge or underhand tricks, but it's not completely clear that the experiment really maps onto a real-world situation in which someone has an impossibly powerful AI in a box and isn't sure whether to let it out, where the stakes are much higher. (Although the impossibly powerful AI in a box would be, as mentioned, *much more convincing* than Yudkowsky, and also able to offer much greater rewards.) But I do find, as Bostrom says, that it undermines my confidence. The use of 'oracle' AIs – superintelligences locked in boxes like this, which are limited to answering questions by text – might well be safer, but it seems optimistic to assume that they couldn't talk us into letting them out of the box.

Chapter 16

Dreamed of in your philosophy

Toby Walsh, the University of New South Wales professor of AI whom I spoke to, is extremely worried about AI safety. But he thinks that the paperclip maximiser scenario is the wrong thing to be concerned about.

'The scenario that I wrestle with,' he told me, 'is 3D-printed drones. You could fill a couple of trucks with these, ride them into New York City and say, "Kill every white person who's here." That could be their code.' This is not – quite – possible with real-world technology right now, he says, but it probably will be in only a few years' time. Autonomous weapons are weapons of mass destruction, he says, and should be outlawed like atomic, biological and chemical weapons are, by international treaties; he has led efforts to lobby the UN to issue a ban. He is concerned about a new 'arms race' as countries rush to build them: once you have autonomous weapons, human-operated weapons will be too slow to be of any use. Russia has an autonomous tank; in the DMZ between North and South Korea, he said, there are autonomous machine guns which will fire on anything human-shaped, and will kill you 'from four kilometres away with deadly accuracy'. The technology for the sort of unstoppable drone swarm he fears is 'not 20 years away. It's 10 years away. Maybe five.' But he is – not dismissive, exactly, of all the LessWrong apocalypse stuff, but certainly wary. 'Most people who believe in the singularity aren't AI researchers,' he told me. 'They're philosophers. Researchers try to build the machines, and therefore appreciate some of the challenges.'

I also emailed Rodney Brooks, another professor of AI, at MIT, asking to speak to him, because I knew he was highly sceptical of all this stuff, and he emailed back in an entertainingly grumpy fashion : 'I regard the people making the claims [about AI risk], and [writing] books like yours, as "flat Earthers". They are 'completely and totally wrong'. 'Focusing on AI diverts attention and wastes everyone's time,' he said, 'while more immediate dangers abound. The destruction of democracy, the end of privacy, the subjugation of the masses to the few – all technology-based disasters.' He felt that claims like Yudkowsky's, that we face an existential threat from AI, 'have no basis', and this message 'gets amplified by other people (e.g. you) writing about the baseless claims'. 'A hype chain reaction has gone off,' he added. 'Each one of you begets 10 more of you.' (I should say that Professor Brooks' grumpiness was very self-aware and, I think, humorous. He signed off 'Curmudgeonly yours, Rodney Brooks'. I didn't get the impression that he was being rude, although nor did I get the impression that he would care a great deal if I *did* get the impression that he was being rude.)

I think saying that concerns about this stuff are the preserve of cranks and philosophers is not entirely untrue, but not really fair, either. I spoke to Toby Ord of Oxford's FHI – one of Bostrom's close colleagues and someone who has spent a lot of time thinking about existential risks. (You met him in the Introduction; he puts AI as one of the top two risks, alongside bioengineered viruses.) He's also a philosopher, so maybe I should declare that as an interest, but he took significant umbrage at the claim that AI risk is just being pushed by philosophers; it was the only time in our long conversation that he became anything other than cheerful and calm. That claim is either 'disingenuous or extremely ignorant', he said. 'It's hard to not be one or the other. I don't want to be too harsh on [people making that claim]: it may not be quite disingenuous, but it's plainly false. You're forgetting Stuart Russell, for example.' (Russell, who is a UCLA Berkeley AI professor, co-authored *Artificial Intelligence: A Modern Approach*, which we've mentioned before. He's on record – many times – arguing that AI risk should be taken seriously.)

'This is actually why I like the survey that Nick Bostrom and Vincent Muller did,' said Ord, which found that AI researchers, on average, think there's a 10 per cent chance that AGI will arrive by 2022; a 50 per cent chance by 2040; and a 90 per cent chance by

2075.[1] And 18 per cent of respondents thought the impact would be 'extremely bad (existential catastrophe)', i.e. human extinction. 'The point isn't that it's 18 per cent exactly,' said Ord. 'If we ran a survey again, we might get 17 per cent or 19 per cent or 20 per cent. The point is that it's not 1 per cent. It is the case that typical AI researchers think there's a very serious chance that their work is going to have immensely negative effects.'

He pointed to a few more people. I'd mentioned Slate Star Codex, the Rationalist Scott Alexander's blog, earlier in the conversation. 'Scott has a very good blog post listing all the AI researchers who think it's a serious problem,' Ord told me, which indeed he does: it's called 'AI researchers on AI risk'[2] and was published in 2015. 'It's a very long list, including, for example, [DeepMind co-founders] Demis Hassabis and Shane Legg. The world's biggest AI company is run by people who think this is a real concern.'

There are several other high-profile names on the list, including Murray Shanahan. I asked him about it as well, and he said – with a lot of caveats – that AI risk really is a thing. 'My view is probably not that far from the highly nuanced Nick Bostrom view, which is that there's an important argument here that deserves to be taken seriously.' His caveats were that it was probably quite a long time in the future, and may not happen at all. He was wary of oversimplification – 'Elon Musk or someone drops a few tweets or soundbites in the media, and all of the nuances and hedges are totally and utterly lost' – but he absolutely agreed that it is a 'realistic issue'. 'Even if it's some time in the distant future, and with a low probability, the possibility of a very bad outcome means we still need to think hard about it,' he said.

I also asked Nick Bostrom, who is very much a philosopher too. And he said that at one point it was mainly philosophers who were worrying about AI risk, and that that probably wasn't a bad thing: 'At the very earliest stages, there is a certain need for conceptual work to be done, when it's not clear yet what the problem is, even, or what the concepts are. You could think that superintelligence could be big, could be powerful, maybe it could be dangerous. But then how do you go from there to actually writing technical papers with math? How do you make progress on that?'

You need to break down this big, difficult question into its constituent parts, he said, and that's where philosophy is useful, 'as one discipline among others'. But 'the further along we go, the greater the relative weight of computer science and mathematics, and that is what we now see happening'. For instance, Bostrom's FHI now does technical seminars with people at DeepMind. 'Every month we have them coming up here, or we go down there, writing papers together and so forth,' he said. 'With MIRI as well, and OpenAI, and there's a group at Berkeley and another at Montreal. It is becoming more integrated with mainstream machine learning. I still think there is scope for more conceptual work to be done, but I think it does as we move closer become more continuous with machine-learning research and computer-science research.'

I suspect that a lot of the arguments come down to a sort of difference in emphasis, rather than major differences in belief. Shanahan said something similar: 'Rod Brooks might say, "This isn't going to happen, this is absurd, it's not going to happen for a hundred years," but if you talk to Nick Bostrom, he'd say, "It's really important that we think about this, because it might happen in only a hundred years."' This is a point that Alexander makes in his blog as well: 'The "sceptic" position seems to be that, although we should probably get a couple of bright people to start working on preliminary aspects of the problem, we shouldn't panic or start trying to ban AI research. The "believers", meanwhile, insist that although we shouldn't panic or start trying to ban AI research, we should probably get a couple of bright people to start working on preliminary aspects of the problem.'[3]

There do exist people like Brooks, who think it is ridiculous. And there are people like Toby Walsh, who worry very much about AI safety but who reckon that this is the wrong sort of AI safety to worry about. But I reason, cautiously, that it is fair to say that AI researchers don't, as a body, regard it as stupid to worry about all this; a significant minority of them believe that there is a non-negligible chance that it could really mess things up. It's not just a bunch of philosophers sitting around in Oxford senior common rooms pontificating.

Chapter 17

'It's like 100 per cent confident this is an ostrich'

Having read all this, you would still be entirely forgiven for thinking it is a bit angels-on-the-head-of-a-pin. We're imagining how a sort of godlike robot-brain in the unspecified-time-from-now future might decide to behave.

But there are, right now, some little hints that AIs are going wrong in ways that are quite paperclip-maximiserish, albeit on a much smaller scale. A great paper was released on ArXiv[1] in March 2018, about digital evolution – machine-learning programs which come up with solutions to problems by mutation and selection in the same way that actual biological life evolved. Typically, there's a little 3D virtual world, a little virtual avatar, and a task of some form, say: 'Travel from location X to location Y.' Then an algorithm is allowed to redesign the avatar using evolution; it copies it however many hundreds of times, each one slightly, randomly different (number of legs, size of feet, etc.), and then lets it try to complete the task. The avatars that are most successful – which walk fastest from X to Y – are then copied again, once more with random variations. And again and again. It's exactly the same process – replication, variation and competition – as in real evolution.

The ArXiv paper was essentially a series of anecdotes from machine-learning researchers, about how they'd done exactly this and found that their little avatars had gamed the system. One was doing the sort of 'walk from X to Y' task we just talked about; it was trying to find innovative methods of locomotion. But the evolution algorithm just found that it was easier to design an avatar that was really, really tall:

basically, a big rope with a weight on the top. When the simulation began, the avatar would just fall over in the direction of Y.

Another experiment involved trying to breed creatures to jump high, by giving them a task of 'get your centre of gravity as high in the air as possible'. But again, the evolution algorithm found a loophole: it just built tall, thin, static towers with a heavy block on the end. The researchers tried to fix it by saying, 'OK, your task is to get the block that started closest to the ground as far as possible from the ground.' So the algorithm stuck with a tall, thin stick with a block on the end, but now just made it do a somersault, so its 'foot' ended up in the air.

At least those ones *sort* of found a solution to the problem that was set. They would be of no use in the development of a walking or jumping robot, but at least the little avatars did get from X to Y, or end up with a centre of gravity high in the air. It's the other ones that are particularly funny-in-a-scary-way when you consider the context.

The algorithm GenProg, short for genetic programming, was given a task of fixing bugs in software. The software it had fixed was then run through a series of tests to see if it worked. The more tests it passed, the more offspring it had. In theory, that would lead to the evolution of better bug fixers. But GenProg found simpler solutions. One was that it was supposed to repair a sorting algorithm that was buggy; it was sorting lists into the wrong order. After GenProg 'fixed' it, the sorting algorithm was run through the battery of tests, and scored perfectly in each one: not a single list was out of order.

But when the (human) programmers went to check, they noticed that GenProg had, instead of fixing the program, just broken it completely. That made the program return an empty list, and an empty list can't be out of order. Sorted! In another experiment, it was supposed to create text files which were as similar as possible to some *target* text files. 'After several generations, suddenly and strangely, *many* perfectly fit solutions appeared, seemingly out of nowhere,' the authors write. It turned out that one of the evolved versions had just deleted the target files, so it – and, subsequently, many other versions – could just hand in an empty sheet and get a perfect score.

And, amazingly, one learned to win a computer Tic-Tac-Toe (noughts and crosses) tournament by forcing its opponents to crash. It tried to make impossible moves on imaginary points on the board, billions of squares away from the actual board. Doing that forced the other programs to try to represent a board billions of squares across in their memory; their memory wasn't big enough, so they crashed, and the cheating algorithm won the game by default.

I spoke to Holden Karnofsky of OpenPhil about AI risk, at the office the organisation shares with its parent, the charity evaluator GiveWell, on the fourteenth floor of a downtown San Francisco skyscraper with an extraordinary view out over the water, and he gave me another example of AI going wrong, right now, in unexpected but related ways.

OpenPhil is the largest single supporter of Yudkowsky's MIRI, across the harbour in Berkeley. It is a significantly slicker-seeming operation, though: it has that Google-style Bay Area corporate feel to it, if not quite as primary-coloured and futuristic. Its offices have confusing names: 'Sesame Street', 'Deworming', 'Cage-Free Eggs' (it turns out that these are examples of how philanthropy has changed the world). I was in Deworming, unsure what message that was meant to send me, watching ferries ply their way across the bay.

Holden spoke to me for a while about ways in which AI can go wrong. One of the big breakthroughs in recent years has been image recognition: you'll have noticed that suddenly your phone can sort your photos, with surprising success, into 'pictures of you' and 'pictures of your husband' and 'pictures of your children'. The face stuff has been particularly impressive, to my mind, but AI can recognise more and more things. Even 10 years ago I remember people talking about how difficult it was to get an AI to tell a dog from a cat, say.

But now, they're amazingly good. And that's because they've been trained on hundreds of thousands of images of different things. But it's been discovered that they can go wrong in surprising ways which, if there was more at stake than a misidentified image, might not be discovered until it was too late.

'A normal image classifier would look at this image,' Holden said, 'and it would say, "That's a panda, I'm 57 per cent confident."' The picture he shows me is, indeed, that of a panda, from a 2015 paper by some Google AI researchers.[2] 'But when you show it this picture, it says, "It's a gibbon," and it's 99 per cent confident.' The picture he then showed me was – as far as I could tell, in a full minute of staring at it – the same picture of the same panda. 'You can't even see the difference,' he said. 'The difference is this.' He held up a picture of what looked like static, the sort you used to get on TV screens when they were tuned to dead channels, except in colour instead of black and white. Each pixel represented a tiny shift in colour between the two images. He showed me another image, of a bus, which the AI had declared with 99 per cent confidence to be a picture of an ostrich. It was honestly quite funny.

What was going on was that the AI had learned to recognise the images by looking at lots of pictures of ostriches, or buses, or pandas, or whatever. And it had pulled out common features of each. But the common features were not the ones humans would recognise: wheel-arches, or long necks, or black-and-white fur. They were weirdly specific. It worked well on naturally occurring images, but could easily be thrown off – amusingly, catastrophically off – by 'adversarial images', images intended to deceive.

'It's learned by training in a very narrow literal way,' said Holden. 'If you try to screw with it you can screw with it very easily.' The static-y pictures were carefully generated to throw the AI off by changing tiny aspects of the image to bring up the weird, specific aspects that it thought of as 'ostrich' rather than 'bus'. It's an example of how an AI with an apparently simple goal ('learn to recognise ostriches') can go wrong in ways that you might not notice until it's too late. 'The concern this raises is that an AI could train on a dataset and when it sees new things, as long as they're similar to the things it's already seen, it's fine. And if they're different it can completely break down in utterly ridiculous ways.' He gestured at the bus. 'It's like 100 per cent confident this is an ostrich.'

OpenPhil is funding efforts to find ways of stopping these sorts of problems – 'trying to get them to not classify buses as ostriches no matter what we do with the bus'. 'We can see we have problems,' he said. 'So we're going to try to solve the problems in the toys like this, and maybe it'll make it safer later. Is it going to make a huge difference? I don't know. But does it seem like it might? Yes.'

But it's an example of something that could go wrong in strange ways – long before human-level AI, when an AI is in charge of, say, self-driving cars. 'It's not a situation you want to be in if, say, an AI is managing the power grid and something weird happens. Where there's a little bit of space that wasn't in the training set and it just totally melts down. Basically, if you imagine a future where you give an AI a goal and it'll maximise the hell out of that goal, but you better hope that you specify that goal perfectly and nothing weird happens, it doesn't see anything it didn't see in its training – in that world I think you would be right to feel very scared.'

Part Three

The Ways of Bayes

Chapter 18

What is rationality?

In writing the Sequences, Yudkowsky's goal, essentially, was to demonstrate that an AI could be intelligent (or rational, or good at 'optimising', or whatever you want to call it) without being remotely like a human intelligence. That required two things. First, he had to explain what rationality (or intelligence, or optimisation power) is. Second, he had to demonstrate how and why human intelligence is idiosyncratic and partially irrational, in order to show that it isn't a template for *all* intelligences.

For Yudkowsky, intelligence/rationality is about matching your mental model of the world to the real world as closely as possible, and about making decisions that achieve what you want them to as often as possible. Both of these processes, he says, can be described using a simple equation called 'Bayes' theorem'. Here's how all that works.

First, we should discuss what Yudkowsky means by 'rational'. There are two fundamental ideas underpinning 'rationality' as defined by the Rationalists. They are 'epistemic rationality' and 'instrumental rationality'.

'Epistemic rationality' is achieving true beliefs. Or, as Yudkowsky puts it, 'systematically improving the accuracy of your beliefs'.[1] The Rationalists have a phrase for this: 'The map is not the territory.' Your mind contains thousands of models, which it uses to predict reality. For instance, I have a working model of gravity and air resistance and things which allows me (sometimes) to catch a ball that is thrown to me. Even more prosaically, I have a model which says, 'The lamp is over there' and 'The door is behind me' and 'The window is in front of me.' The degree to which I have an accurate model, the degree to which I can walk to where I think the door is and actually find a door there, is the degree to which my model corresponds with the world, or my 'map' corresponds with the 'territory'. 'This correspondence between belief and reality is commonly called "truth", says Yudkowsky, 'and I'm happy to call it that.'[2]

I've used prosaic examples, but it applies just as much to more abstract ones. Whether or not black holes emit Hawking radiation is a question of fact. If your model predicts, as the late Stephen Hawking's (complex, mathematical) model did, that black holes *do* give off radiation, then that is a statement about your mind. Whether your model is correct does not depend on how convincing your arguments for Hawking radiation are, or how strongly you believe it, but on whether, out there in the universe, black holes really do give off radiation. The only way to see whether your map corresponds with the territory is to go and look, or otherwise seek evidence. Insofar as the universe behaves in ways that your model predicts, your model is good; insofar as it doesn't, it isn't.

Instrumental rationality, by contrast, is about your actions. 'Rationalists', says Yudkowsky, 'should win.'[3] The idea is the same as the definition of 'behaving rationally' in the textbook *Artificial Intelligence: A Modern Approach*, which we discussed in the Introduction. It is *choosing that course of action which is most likely, given what you know now, to achieve the goal you want to achieve.* It doesn't mean, he says, selfish domination, or money, or anything specific. It means 'steering reality – sending the future where you want it to go'. That could mean to your own selfish ends, or it could mean towards preventing climate change, or turning the universe into paperclips. It is about *successfully doing what you wanted to do*.

There are various corroborating mathematical ideas which we'll come to in due course. But this is the most fundamental thing. Yudkowsky refers to the semi-legendary Japanese swordmaster Miyamoto Musashi, who said of his art: 'You can win with a long weapon, and yet you can also win with a short weapon. In short, the Way of the Ichi school is the spirit of winning, whatever the weapon and whatever its size.'[4]

Instrumental rationality doesn't, necessarily, mean behaving in a 'rational' way, as defined by Hollywood and especially Mr Spock. Yudkowsky really doesn't like Spock. 'Consider Mr Spock of *Star Trek*, a naive archetype of rationality,' he grumbles at one point. 'Spock's emotional state is always set to "calm", even when wildly inappropriate.' If you are about to be blown up by a Klingon torpedo, then being afraid might be rational. Worse than that, Spock's 'rational' predictions, given in spuriously precise percentages, are usually wrong. 'He often gives many significant digits for probabilities that are grossly uncalibrated,' says Yudkowsky. 'E.g.: "Captain, if you steer the *Enterprise* directly into that black hole, our probability of surviving is only 2.234 per cent." Yet nine times out of ten the *Enterprise* is not destroyed. What kind of tragic fool gives four significant digits for a figure that is off by two orders of magnitude?'[5]

Instead it means winning.

Let's return to Newcomb's paradox for a moment, in which a superintelligent alien AI, Omega, comes to Earth and offers you two boxes, one transparent and with £1,000 in it, the other opaque and containing either £1,000,000 or nothing. Robert Nozick, the great American philosopher, wrote an essay on this problem which is notable, among philosophical essays, for its endearingly baffled tone. 'I have put this problem to a large number of people, both friends and students in class,' he writes. 'To almost everyone it is perfectly clear and obvious what should be done. The difficulty is that these people seem to divide almost evenly on the problem, with large numbers thinking that the opposing half is just being silly.'[6] At the end of his first section he asks people to stop reading to think about it themselves: 'It is not that I claim to solve the problem, and do not wish you to miss the joy of puzzling over an unsolved problem,' he says. 'It is that I want you to understand my thrashing about.'

On the one hand, it's obvious that you should 'one-box'. You've seen Omega being right before. It is highly likely that it has predicted your decision correctly. If someone were watching you, and betting what you'd get if you picked both boxes, they would bet confidently and at high odds that you would only get £1,000. Worse than that, if you picked both boxes, and before the results were revealed were asked to bet what you will get, you would rationally bet that you would only get £1,000. 'Knowing all this,' says Nozick, 'do you really want to take what is in both boxes, acting against what you would rationally want to bet on?'

But. But, but but. It's obvious that you should 'two-box'! Imagine that the far side of the £1,000,000-or-nothing box is transparent, and a friend of yours can see what's in it, and is watching you make your decision. She's been staring at this box for an hour, she can see there's £1,000,000 in it, or nothing in it. *Whatever is in the box*, she will be hoping that you take both! If there's nothing in it, pick them both: you'll at least get £1,000. If there's £1,000,000, pick both: you'll get £1,001,000. The time at which Omega could affect it has passed! It's already flown off to its home planet!

Spock, I think, would two-box. It is not logical, Captain! The being has flown away. The boxes contain what they contain. And Nozick, after endless agonising and vast scrawls of mathematical notation which I can't follow, says he would two-box.

Yudkowsky and the Rationalists would say: one-box. (I don't know if Kirk would one-box, but Kirk seems to win even when Spock tells him he won't. Certainly, Kirk is the better Rationalist than Spock, even though Spock is the 'rational' one.) There are mathematical reasons for this: an entire branch of decision theory which Yudkowsky has developed, in fact. But the underlying reason is that *people who choose one box make more money than the ones who choose two*. And that, assuming that you would rather have more money than less, is the whole thing. That's the game. For one-boxers, says Yudkowsky, this 'is a simple dilemma and anyone who comes up with an elaborate reason why it is "rational" to take both boxes is just outwitting themselves. The "rational" chooser is the one with a million dollars.'[7]

That's the Rationalist rationality at its most basic, then: trying to believe things that are true, and trying to take decisions that lead to the outcomes you want. Obviously, it all sounds a bit un-derwhelming when I put it like that. We should go into it in a bit more depth.

Chapter 19

Bayes' theorem and optimisation

For Yudkowsky, the heart of rational behaviour is the simple mathematical equation known as Bayes' theorem. When he talks about rationality, he is talking about Bayes; the project of improving human rationality is a project of making humans better Bayesians. The theorem is (he says, and decision theory agrees) absolutely central to what good decision-making involves. When you have evidence for something, that evidence allows you to shift your beliefs only as far – no more, no less – as the distance dictated by Bayes.

The Reverend Thomas Bayes is buried, rather appropriately, a few hundred yards from the offices of the Royal Statistical Society, in Shoreditch, east London. Bayes, a somewhat obscure eighteenth-century Presbyterian minister with a sideline in mathematics, wrote a couple of well-received books in his lifetime, one on theology, the other defending an aspect of Newton's calculus from criticism by George Berkeley. The latter, according to Wikipedia at least,[1] appears to have been enough to have got him elected a Fellow of the Royal Society.

Neither of these books are what he is remembered for. In later life he became interested in probability, and after he died his friend Richard Price edited his notes on the subject into an essay for the journal *Philosophical Transactions*.[2] It was called 'An Essay towards solving a Problem in the Doctrine of Chances', and contained within it a simple equation which underlies the whole of modern probability theory.

Bayes' theorem goes: $P(A|B) = (P(B|A)P(A))/P(B)$. Don't worry if you can't follow the notation, it doesn't matter. It's really very easy to understand. It's working out how likely it is that statement

A is true in the event that statement B is true. In full, it says that the probability of A given B equals the probability of B given A, multiplied by the probability of A on its own, divided by the probability of B on its own.

That probably didn't help. But honestly, it's very easy to understand. If we move it away from the realm of abstract letters, it'll become clearer. Imagine you've got a blood test that screens for cancer. Let's say that 99 per cent of the time, if you have cancer, it will tell you, correctly, that you have cancer. And 95 per cent of the time you test someone *without* cancer, it says, correctly, that they don't have cancer. Knowing that, if the blood test comes back positive, what is the likelihood that you have cancer? It's about 95 per cent, right?

No! The answer is *you have absolutely no idea*. There is not enough information given to provide you with the faintest clue what your chances of having cancer are. That's because, without knowing how common the cancer you're looking for is, you don't know how common false positives will be.

Let's say one person in every 1,000 in the population has this cancer at any given time, and you run your test on a million people. On average, there will be 1,000 people among those million who actually have cancer. Your test will correctly identify that 990 of them have cancer. Of the 999,000 people who *don't* have cancer, 5 per cent of them will be told that they do. That's 49,950 people. So you'd get 50,940 positive results, but only 990 of those would actually have cancer. If you go into the clinic, and have a test that the doctor says (truthfully) is 95 per cent accurate, and it comes back positive for cancer, then, in this case, your chances of actually *having* cancer would be rather less than 2 per cent. (And, of course, 10 people will go happily home thinking they are healthy, but will in fact have cancer.)

The background rate of cancers in the population is, when we're talking about Bayes' theorem, your *prior probability rate*. Any new information pointing you towards some conclusion – say, a positive cancer test – is only useful in the light of your prior assessment of how likely that conclusion is. This is, to put it mildly, not obvious. If you hear that a test is 95 per cent accurate, it seems reasonable to assume that if it gives a positive result, there's about a 95 per cent chance that it's right. But that's not true at all.

If you find this counterintuitive, don't worry. So does everyone else. If anyone should get this stuff right, it should be doctors, who make decisions on the basis of cancer tests and background rates all the time. But, as Yudkowsky points out, they don't.[3] In one experiment, whose findings have been replicated several times, less than one doctor in five gave the right answer to a similar question; nearly half of them said 95 per cent, and the average guess was 55 per cent, a 30-fold overestimate of the true answer.[4]

For Yudkowsky and the Rationalists, Bayes' theorem essentially *is* rationality: every decision-making process is successful insofar as it emulates Bayes. 'Eliezer's position is that every successful process owes its success to "moving in harmony with the Bayes", as he'd say,' comments Paul Crowley. 'That probability theory says that this is the only place success comes from.' Any process which moves steadily closer towards true answers and successful decisions must, says Yudkowsky, be doing so in a Bayesian way. In Yudkowsky's own words: '[If] a mind is arriving at true beliefs, and we assume that the Second Law of Thermodynamics has not been violated, that mind must be doing something at least *vaguely* Bayesian – at least one process with a sort-of Bayesian structure somewhere – or *it couldn't possibly work.*'[5]

I asked him about this, and he compared it to the laws of thermodynamics. There's a theoretical device called a Carnot engine, an idealised version of a heat engine – something like an internal combustion engine or a steam engine, something which uses thermal energy to do mechanical work. According to the laws of thermodynamics, if you had a perfectly efficient heat engine, there is a maximum amount of work you can get out of a given amount of energy. 'Cars don't run on ideal engines, but they can't violate thermodynamics,' says Yudkowsky: any real-world engine must be less efficient than the perfect Carnot thought experiment.

He says this is analogous to how Bayes' theorem relates to decision-making: it is *ideal* decision-making. It is what decision-making looks like when it happens impossibly perfectly. 'This is a point that confuses people who think in terms of a toolbox and think Bayesian methods are just one more tool in the box,' says Yudkowsky. It's not. All the other things in the decision-making toolbox are useful insofar as they approximate the Bayesian equation. Any process that uncovers true facts or makes good decisions is doing something Bayesian, whether it's evolution or human thought or anything else. Anything that looks like decision-making must run on something like Bayes. 'To the extent that [decision-making processes] work at all, they must necessarily work because they have bits of Bayesian structure embedded in them.'

In the Carnot-engine analogy, 'Bayes is akin to the laws of thermodynamics,' Yudkowsky says, 'and a little program that directly implements Bayes' rule is like a thermodynamically ideal Carnot engine.' Just as no car can ever really run on a Carnot engine – energy will always be lost into the world – so no decision-making system can ever be a perfect Bayesian one. 'The algorithm that uses the work perfectly is too expensive to implement,' says Yudkowsky. 'But some of that work must be performed somewhere, or the "car" doesn't move at all.'

So, for instance, evolution. The astronomer Fred Hoyle used to say that the chances of evolution producing life were like those of a whirlwind passing through a junkyard and producing a Boeing 747.[6] The number of ways in which the components of a 747 can be arranged are unimaginably huge, and only a tiny, tiny fraction of a fraction of a percent of them would be able to fly; likewise, if you took apart, say, a lemur, down to its constituent cells, and put it back together again at random, you would be vanishingly unlikely to create with any great success something that swung through trees.

But, of course, 747s aren't made by whirlwinds, they're designed by humans, who are good at looking at a pile of aerospace equipment and immediately ruling out the overwhelming majority of possible combinations. And, despite Hoyle's misunderstanding, lemurs aren't made at random, but by the slow, inefficient but still non-random process of evolution by natural selection.

Yudkowsky describes both human intelligence and evolution as *optimisation processes*. An optimisation process is a way of moving through a huge space of possibilities to get close to, and hit, the target that you want.

And that's what the Bayesian equations are doing. With the cancer test we discussed above, the original space is quite large – a million people who may have cancer – and the target is quite small – 1,000 people who actually have cancer. Your amazingly accurate cancer test does not magically provide you with true positives, but it allows you to narrow your space. Your prior probability of any random person you grab having cancer is one in 1,000, or 0.001. You do your test, you eliminate about 994,000 possibilities; the chance of any given person is now about 0.02. You've narrowed your space down to about one-twentieth of the size it was. If you were to do the test again (assuming the false positives were random, not systematic), then you'd narrow it down still further. You are optimising your search, incrementally closing in on the truth. The phrase that gets used a lot is 'probability mass': what weight of probability do you assign to each possible outcome? Before you did your cancer test, you put only 0.1 per cent of your probability mass on the outcome 'cancer' and 99.9 per cent on the outcome 'no cancer'. The test allows you to shift some of the mass; you now put 2 per cent of your probability mass on 'cancer' and only 98 per cent on 'no cancer'. Your probability mass always has to add up to 1, to a 100 per cent chance; each new piece of evidence just needs to shift it around between options.

Human intelligence does the same thing. If you took all the parts of a 747 and put them together at random, the chance of any one of the planes flying would be – I don't know. Ridiculously tiny. Something on the order of millions of zeroes after the decimal point, I expect, depending on what counts as a 'part' of a 747: if you take one figure I heard, that a 747 has 6 million parts, then Wolfram Alpha says that the number '6 million to the power 6 million' has around 40 million digits. The probability mass I would assign to 'spontaneous generation of a 747' is unimaginably small.

But if you said to a *human*, 'Arrange those parts in such a way as they might fly', your odds would go up immensely. Even if it were me, I'd know to do things like put any large, flat panels in wing-like configurations on the side. I'm sure my chances of making a flying thing out of a junkyard full of aero parts would be, gosh, at least 1 in 100 million. If the person putting the parts together were an engineer, rather than a glorified typist, that chance would drop even further; if she were an aeronautical engineer it might even reach close to evens. And if that engineer is allowed to have several goes, and test each version, and incrementally improve it, and ask for help and read books and so on, then it would become a near-certainty.

It's the same with evolution. The chance of any random arrangement of organic molecules forming a living, breathing animal, or any complex creature, is extremely small: there is a huge number of possible arrangements, the 'space' you are searching in, and only a microscopically tiny fraction of those will 'work'.

But if you already *have* a simple, self-replicating thing, which makes slightly imperfect copies of itself, then it will start to 'search' that space. The copies which are worse at replicating will tend to be eliminated; the copies which happen to be better will tend to spread. The diverging types will tend, simply by random movement being pruned by non-random selection, to move towards those 'areas' of the 'space' which represent functioning organisms. Your probability of finding a working organism if you throw together a random agglomeration of organic parts is essentially zero; your probability of finding one after a few million years of evolution from a simple replicator is close to 1.

(Note: you still need to *have* a simple replicator, created by some random or at least natural process. Evolution doesn't explain the *very start* of life. There are various ways it might have arisen: Nick Lane, a professor of biochemistry at UCL, thinks that very simple cells with the right sort of chemical and energy gradients could have formed naturally in vents at the bottom of the sea. But whatever it was, it only had to happen once, in millions of years, across a whole planet. Billion-to-one chances aren't unlikely if you have a billion chances. 'Since the beginning', goes one Rationalist haiku, 'not one unusual thing / has ever happened'.)

Bayes' theorem is extremely useful from a philosophical point of view. I studied philosophy at university, and there were endless arguments about the 'problem of induction'. The idea was that you could see a million white swans, but you would never be able to prove the statement 'all swans are white', because it would take seeing just one swan which was black – which Western explorers did when they first reached Australia – to disprove it. No amount of 'inductive reasoning' – coming to conclusions from evidence – could ever prove anything.

But Bayesian thinking lets you sidestep this altogether. You simply learn to think probabilistically. Having never seen a swan, you might assign a prior probability to the hypothesis 'all swans are white' of, say, 1 per cent. (All swans could be green, for all you know.) You see your first swan, you update your prior probability in the light of new evidence: you might think that it's now 15 per cent likely that all swans are white. (You've only seen one swan. They could come in all sorts of colours.) That is now your new prior.

But after wandering around Renaissance Europe for 40 years, only ever seeing white swans, and constantly updating your priors, you are now much more confident in the statement. As a good Bayesian, you're never certain, but you've seen thousands of swans, each one adding a small dollop of evidence to support your hypothesis, so you push your confidence up to a very solid 95 per cent.

Then you get on a boat to Van Diemen's Land, and you see a black swan. Your confidence immediately plummets to 0.01 per cent. The problem of induction isn't a problem any more, as long as you're willing to think in terms of likelihoods and probabilities, rather than certainties. You're never certain – someone might be painting all those black swans black, or you might be hallucinating – but the more swans you see, the more you can update your priors and increase your confidence. The Rationalists think of all knowledge in these terms: how confident you can be in your beliefs, how much 'probability mass' you should assign to some proposition, and how much you can 'update' your beliefs in the light of new evidence.

There's another philosophical problem, the 'paradox of the ravens', which you can also solve with Bayesian reasoning. It starts similarly. The statement 'all ravens are black' is logically equivalent to 'if something is not black, it is not a raven'. That's because anything that renders the first statement untrue would also render the second one untrue, and vice versa. But that leads to a strange situation. Seeing a black raven should count as evidence for the statement 'all ravens are black'. But if that's true, then seeing a non-black non-raven (say, a purple hat) counts as evidence for the statement 'if something is not black, it is not a raven'. And if *that's* true, and the statements 'all ravens are black' and 'all non-black things are non-ravens' are equivalent, then seeing a purple hat is apparently evidence that all ravens are black.

Again, this has been argued about for years; the thought experiment was first proposed in the 1940s. But with Bayesian thinking, it's nice and straightforward. The purple hat is indeed evidence. It's just not very much evidence. You adjust your prior an infinitesimal amount and carry on. Absence of evidence is, in fact, evidence of absence, even if not strong evidence. (I should say: this isn't something the Rationalists came up with. Bayesian solutions to the problem of induction and the paradox of the ravens are decades old. And people still get very angry about them and argue that they're wrong. They're not *the final word*. But, to my mathematically ungifted mind at least, they seem to provide commonsensical ways around the problems.)

As we saw at the beginning of the chapter, for Yudkowsky – and for decision theorists such as E.T. James, author of *Probability Theory: The Logic of Science*, whom Yudkowsky frequently cites as an inspiration – the Bayesian equation is the iron law of decision-making. There is a correct amount by which you should shift your beliefs in the light of new evidence. Shifting your beliefs by more or less is simply wrong.

An example that Yudkowsky uses in the Sequences is the lottery.[7] Imagine a box, he says, that beeps every time you choose a winning lottery ticket. It'd be no use whatsoever if it also beeped every time you chose a *losing* ticket, obviously. But it doesn't – it only beeps 25 per cent of the time when you have a losing one. So, in a lottery with six numbers and 70 balls, there are 131,115,985 possibilities. You write down your lottery numbers. The machine beeps. What are your odds?

Well, it'll beep on the correct one, but it'll also beep on one-quarter of the 131,115,984 other ones, leading to an average total of 32,778,996 false positives. Your 75-per-cent-accurate test lets you move from 1/131,115,985 to 1/32,778,996. That is how much you can update your beliefs. Any more and you're overconfident, any less and you're underconfident.

'You cannot defy the rules,' writes Yudkowsky. 'You cannot form accurate beliefs based on inadequate evidence. Let's say you've got 10 boxes lined up in a row, and you start punching combinations into the boxes. You cannot stop on the first combination that gets beeps from all 10 boxes, saying, "But the odds of that happening for a losing combination are a million to one! I'll just ignore those ivory-tower Bayesian rules and stop here." On average, 131 losing tickets will pass such a test for every winner.'[8]

This is a toy example, obviously. It's rare that you find situations that are so neatly mathematically defined in real life. You'll have to use a lot of guesswork and intuition. But even when you can't measure the odds so precisely, when you are presented with evidence ('The driver in front of me is driving erratically') for a hypothesis ('The driver in front of me is drunk'), there is a correct amount of confidence that you are allowed to have, based on how common drunk drivers are, how often drunk drivers drive erratically, and how often people drive erratically for other reasons.

Trying to believe on less evidence, Yudkowsky writes, is 'like trying to drive your car without any fuel, because you don't believe in the silly-dilly fuddy-duddy concept that it ought to take fuel to go places. You can try, if that is your whim,' he says. 'You can even shut your eyes and pretend the car is moving. But to really arrive at accurate beliefs requires evidence-fuel, and the further you want to go, the more fuel you need.'

It's not that humans *don't* do something like this – of course we assess evidence and come to beliefs on the basis of that evidence, although usually not by consciously calculating probabilities. (And sometimes we do believe things without evidence, and even celebrate that, a process we call 'faith' and which Yudkowsky would probably call sitting in your car in your driveway, insisting you've driven to Dorset.) But when we change our beliefs in the light of new evidence, even if it feels like an instinctive, gut-level process, we are performing something approximating to Bayes, and we are either getting it right or wrong – we are updating our beliefs the right amount, or too much, or too little.

And, similarly, an AI would be 'intelligent', or rational, insofar as it applied Bayes' rules.

Chapter 20

Utilitarianism: shut up and multiply

So you have your means of updating your beliefs according to how much evidence you get, in Bayes' theorem. But there's another element you need, says Yudkowsky, which is establishing how much something *matters*. You can work out how many people have cancer using your Bayesian test, but in order to make a decision about how much money to spend treating those cancers, you have to think about how much good treating them will do.

This is where it all gets into moral philosophy. And in fact, the thing I like most about the Sequences is that they remind me of one of those enormous eighteenth-century works of philosophy, by Spinoza or Leibniz or someone, that set out to explain pretty much everything. There's this sprawling, ambitious feel to them. The nature of consciousness, the nature of reality, evolution, human psychology, probability, morality. It all feels – to me, at least, and I'm not completely clueless about these things, although it's more than a decade since I did any academic philosophy – like a robustly commonsensical application of modern science to philosophy. David Hume would probably have enjoyed it.

The 'morality' aspect is particularly interesting. Yudkowsky deliberately tries to take it away from the meta-ethical, finding-esoteric-flaws-in-the-ethical-system stuff and into basic numbers. I'll go into what I mean by that in a second, but I think it's most simply expressed in a tweet by someone else entirely, a web developer called Mason Hartman, who was talking about the ethics of self-driving cars:

Philosophy: so sometimes it goes haywire & ends up—
Me: do the thing that kills fewer people

Philosophy: but it's very salient th—
Me: do the thing that kills fewer people
Philosophy: but the human element of control is—
Me: do the thing
Me: that kills
Me: fewer people[1]

(Yudkowsky retweeted it, and I know retweets ≠ endorsement and all that, but I'm pretty sure this one does.)

Essentially, he (and the Rationalists) are thoroughgoing utilitarians. *Do the thing that (you reasonably expect will) kill the fewest people/make the most people happy/cause the least pain.* You can think about it in more detail than that, they would say; but if your thinking pushes you away from doing that, then your thinking has probably gone wrong.

Utilitarianism is the moral philosophy, most associated with Jeremy Bentham and John Stuart Mill, which claims (in Bentham's words) 'that the greatest happiness of the greatest number is the foundation of morals and legislation'.[2] Nowadays most utilitarians don't talk about 'happiness' quite so much. If we naively take 'happiness' at face value, it might be the 'moral' thing to do to plug everyone into machines that artificially stimulate the pleasure centres of our brains, but most of us would not want that done to us, so most utilitarians now talk in terms of 'utility', a sort of code for 'what we want out of life'. Instead of a life of artificially induced bliss, I might prefer a life in which I gain a sense of achievement via actually doing things; modern utilitarianism would award moral points for systems which allowed me that life, instead of sticking me into the pleasure machine.

The Yudkowskian take on it all is admirable, I think, for two reasons. One, it accepts with equanimity one of the really hard conclusions of utilitarianism; two, it gives reasonable, sensible ways around two others. I do not suggest for a second that a series of blog posts written between 2007 and 2009 has answered the big questions of morality that have been batted around by philosophers for 3,000 years, but it fits very neatly with my own intuitions of morality. (That said, I remember in my very first philosophy lecture in my very first year of undergrad at the University of Liverpool in 2001, our head of department, the excellent Professor Stephen Clarke, warned us: in philosophy, you often read an argument and think, 'Yes, I agree with that, that makes complete sense.' Then you read another argument which entirely contradicts the first argument, and you think, 'Yes, I agree with that too.' Be wary of agreeing with things you've just read is, I suppose, the lesson.)

The first problem is the following. A key tenet of utilitarianism is that utility can, in some way, be compared between people. The Rationalists talk in terms of 'utilons', imaginary measures of utility; earlier utilitarian philosophers use the term 'utils'. Obviously you can't really measure them, but you can do thought experiments by putting rough estimates on things: you could imagine that finding £10 on the street is worth one util, say, while getting a job you love is worth 5,000 utils. Giving 5,000 people £10 would then be equivalent to finding someone a job they loved.

But this leads to a difficult situation. Say we imagine something that causes a *huge* loss of utils for one person – something like being horribly tortured for 50 years. And imagine something that causes a tiny, negligible loss of utils – for example, 'suppose a dust speck floated into your eye and irritated it just a little, for a fraction of a second, barely enough to make you notice before you blink and wipe away the dust speck'.[3] If there is anything to this form of utilitarianism, if it means anything at all to say that one experience can be compared to another in some sense, then *some sufficiently large number of people getting dust in their eye is worse than a person being tortured for 50 years.*

Here's a large number. It's a large number that gets thrown around a lot in Rationalist blog posts as a sort of shorthand for 'big. Really big. You just won't believe how vastly, hugely, mind-bogglingly big it is'-type numbers. The number is $3\uparrow\uparrow3$. Here's what that means: $3\uparrow3$ means 'three to the power three', three times itself three times. That's 27. $3\uparrow\uparrow3$ is 'three to the power (three to the power three)', three times itself 27 times. That is 7,625,597,484,987 (getting on for 8 trillion, if you prefer words). $3\uparrow\uparrow\uparrow3$ is ... I lose track a bit, to be honest. Here's Yudkowsky: '$3\uparrow\uparrow\uparrow3$ is an exponential tower of 3s which is 7,625,597,484,987 layers tall. You start with 1; raise 3 to the power of 1 to get 3; raise 3 to the power of 3 to get 27; raise 3 to the power of 27 to get 7,625,597,484,987; raise 3 to the power of 7,625,597,484,987 to get a number much larger than the number of atoms in the universe, but which could still be written down in base 10, on 100 square kilometres of paper; then raise 3 to that power; and continue until you've exponentiated 7,625,597,484,987 times.'[4] This is a *very, very large number.*

So is that number of people having to blink a little worse than someone literally being tortured for 50 years? If you think there is any sense in which utils exist, that experience A can be traded off against experience B, then *surely* a number as enormous as $3\uparrow\uparrow\uparrow3$ is enough to bridge the gap between torture and dust specks. (And if it's not, how about $3\uparrow\uparrow\uparrow\uparrow3$? That is rather a lot bigger.)

Before we go any further, think which you'd pick. Torture? Or an incomprehensibly large number of dust specks?

Yudkowsky's blog post on this, 'Torture vs Dust Specks', was one of the most controversial of all the hundreds in the Sequences. He ends it by saying: 'Would you prefer that one person be horribly tortured for 50 years without hope or rest, or that $3\uparrow\uparrow\uparrow3$ people get dust specks in their eyes? I think the answer is obvious. How about you?' For the avoidance of doubt, his 'obvious' answer is that the dust specks are worse than the torture. In the comments, Robin Hanson agrees, but almost everyone else argues the opposite.

The argument against is that there is no continuity; that you simply can't compare this sort of minor inconvenience to decades of torture. But Yudkowsky argues in a follow-up that this is incoherent. He starts by assuming that we're dealing with a much smaller number of dust specks than $3\uparrow\uparrow\uparrow3$: a googolplex. (A 'googol' is 1 followed by 100 zeroes; a googolplex is 1 followed by a googol zeroes. It is a big number, but much, much smaller than $3\uparrow\uparrow\uparrow3$.) 'Suppose you had to choose between one person being tortured for 50 years, and a googol people being tortured for 49 years, 364 days, 23 hours, 59 minutes and 59 seconds,' he says. 'You would choose one person being tortured for 50 years, I do presume; otherwise I give up on you. And similarly,' he continues, 'if you had to choose between a googol people tortured for 49.9999999 years, and a googol-squared people being tortured for 49.9999998 years, you would pick the former.'

You can carry on doing this, he says. You can keep gradually reducing the *amount of torture per person*, while exponentially increasing *the number of people being tortured*, 'until we choose between a googolplex people getting a dust speck in their eye, and [a googolplex divided by a googol] people getting two dust specks in their eye'.[5] If you think that the former is worse than the latter, then you're committing to the idea that the $3\uparrow\uparrow\uparrow3$ dust specks are worse than the torture, or that there is a sharp discontinuity at some point where, say, 23.6652647 years of torture for one person is worse than 23.6652646 years of torture for a googol people.

I find it difficult to feel, on an intuitive level, that dust specks could add up to torture. But the way I think about it is this. Every tiny little bit of discomfort presumably makes life a tiny little bit less worth living. There is presumably some threshold between 'life worth living' and 'life not worth living'. With a vast number of people like $3\uparrow\uparrow\uparrow3$, or even a mere googolplex, the number of people tipped over that threshold by a minuscule discomfort like a dust speck would be enormous; quadrillions of people, septillions, I literally have no idea except that it would be vast. Again, I don't claim that this is the final word on a problem that utilitarian/ consequentialist philosophers have kicked around for centuries, although I do know there are professional moral philosophers who would argue in favour of choosing the torture over the dust specks.[6] All I can reasonably say is that it fits my intuitions – or, more precisely, that when I follow through the arguments I find that *rejecting* the dust-specks-are-worse-than-torture position violates my intuitions more severely than the alternative.

The second hard-to-swallow endpoint of utilitarianism is what the British philosopher Derek Parfit, who died during the writing of this book, called 'the Repugnant Conclusion'.[7] Imagine you've got a population of a million people living happy lives with loads of resources, says Parfit. Then imagine you add one person whose life is pretty bleak but slightly better than being dead, and redistribute the resources around everyone fairly. By the logic of utilitarianism, you've added utility to the system, so the million-and-one is better than the million, even though the average happiness has gone down slightly.

But then you do it again, and again. Taken to its logical conclusion, a universe containing a trillion (or a googolplex, or $3\uparrow\uparrow\uparrow3$, or however many) lives that are grim and unrewarding and dull but *just about better than being dead* is a better universe, morally speaking, than one with a billion people living extremely rich, fulfilling lives. That feels wrong to me and, I suspect, to most people.

Parfit phrased it like this: 'For any possible population of at least 10 billion people, all with a very high quality of life, there must be some much larger imaginable population whose existence, if other things are equal, would be better even though its members have lives that are barely worth living.'

There have been various attempts to circumvent the Repugnant Conclusion – for instance, arguing that *average* happiness should be taken into account to some degree. Most of the potential solutions have problems of their own and, of course, philosophers have kicked them all around for decades. (I think it is fair to say that every attempt at a coherent philosophical system of ethics leads *eventually* to some awful conclusions, which, as we'll see later, generally involve running one or more people over with a railway trolley.) But Yudkowsky approaches it in a way I hadn't seen before, although I'm sure it's not completely original. He argues that the apparent force of the Repugnant Conclusion comes from its 'equivocating between senses of *barely worth living*'.

'In order to *voluntarily create* a new person,' he writes, 'what we need is a life that is *worth celebrating* or *worth birthing*, one that contains more good than ill and more happiness than sorrow – otherwise we should reject the step where we choose to birth that person.' We should celebrate the birth of a new person we have voluntarily chosen to create: 'Each time we voluntarily add another person to Parfit's world, we have a little celebration and say with honest joy, "Whoopee!", not "Damn, now it's too late to uncreate them."'

If we are saddened to hear the news that a person exists – if their life is sufficiently not-awful that they don't actually want to end it, but still bleak enough for us to feel it is not a joyous thing that they have been born – then we are still obliged to try to take care of them, and improve their lives in such ways as we can. But for *bringing new people into existence*, we should have a higher bar.

'And then the rest of the Repugnant Conclusion - that it's better to have a billion lives slightly worth celebrating, than a million lives very worth celebrating – is just "repugnant" because of standard scope insensitivity [see the next chapter, 'What is a "bias"?']. The brain fails to multiply a billion small birth celebrations to end up with a larger total celebration of life than a million big celebrations.'[8]

I am entirely confident that moral philosophers could dig into this approach and find ways in which it would lead inevitably to stipulating that we torture children or something. (One comment under Yudkowsky's blog post suggested that it could lead to the Sadistic Conclusion, which is that it would be better to create a small number of people living lives *not* worth living than a large number of people whose lives are *just barely* worth living.) But, again, to me it feels like a relatively sane way around the problem.

The third and final problem of utilitarianism is that of 'ends justifying the means'. The classic example is that, if we could cheer up 99 per cent of the population by blaming the remaining 1 per cent for their problems and then imprisoning and torturing that 1 per cent for their imagined crimes, then (assuming that the gain in happiness for the 99 per cent outweighs the loss for the 1 per cent) that would be a moral thing to do. (This is called the 'tyranny of the majority', and John Stuart Mill, one of the first and greatest proponents of utilitarianism, raised a concern about it in his 1859 book *On Liberty*.)

There are lots of other possible examples. The trolley problem is intended to divide people down deontological or utilitarian lines: if you see a railway trolley heading towards five workers on a track, and you can pull a switch so it goes the other way, but there is one person working on that line, should you do it? A utilitarian should, in theory, say 'yes', but a deontologist (someone who follows strict moral rules rather than considering consequences) should say that you never actively kill someone, so you shouldn't pull the switch even though you would save lives. (This is an enormous oversimplification of both deontology and utilitarianism.)

Yudkowsky approaches the trolley problem like this. Sure, he says, it might be the case that you think you can save five lives by killing one. (Or that you can help the poor by robbing a bank, or that you can improve society by staging a military coup and taking over, or any one of 100 versions of 'I can justify Bad Thing X by promising Good Consequence Y'.) But knowing *humans*, it is very unlikely that you are right – or that you are likely enough to be right that, if you did it a million times, you'd overall prevent more harm than you caused.

In the trolley problem, the philosopher stipulates that you know with certainty that your action will save five and kill one and there's no other way around it. But *in reality*, your inadequate human brain can't ever be certain enough that that's the case. You've evolved, as a human, a whole range of systems for creating moral-sounding reasons for doing self-interested things. You are more likely to do good, overall, by implementing the rule 'Never kill anyone' than by trying to work out the maths of utilitarianism on the fly in sudden, stressful situations. And that ends up creating odd-sounding meta-rules, such as 'For the good of the tribe, do not murder even for the good of the tribe.'[9] It is more likely that the thing you think of as being *for the good of the tribe* is in fact for the good of you.

Yudkowsky doesn't broach the specific topic of 'imprisoning and torturing 1 per cent to cheer up the 99 per cent' in a blog post, and since that seemed the most obviously controversial application of utilitarianism, I asked him about it. He replied saying that my numbers were silly. 'Immiserating 1 per cent of the population seems like it would do more than 99 times as much damage to each member of that population as the vague, passing pleasurable thoughts in the 99 per cent,' he said. 'Like, even linearly adding up the pleasure and pain by intensity and number of seconds will say, "No you should not do that".' A better example, he suggested, might be 'asking about immiserating 1,000 people on all of Earth, or one person'.

He suggested that people (particularly and especially me, or at least that was the impression I got) may not be 'smart enough' to try to implement 'utilitarianism' in a way that 'is actually utilitarian'. For a start, we don't tend to run the numbers in the way he mentioned above; we might just hear, 'policy X makes a large number of people happy' and then think, 'therefore utilitarianism demands it', without considering its other effects.

More interestingly, though, he pointed out that 'blaming other people for your problems' isn't, in the wider sense of utilitarianism, necessarily something we'd want. Remember the broader sense, moving away from Bentham's somewhat naive 'greatest happiness for the greatest number', and thinking instead in terms of welfare, or utility, or achieved preferences? I might prefer *not* to be happy, if I knew that my happiness was caused by blaming other people unfairly for my problems. It might not even be, as Yudkowsky puts it, something I 'want to obtain, even for free, by torturing *imaginary* people depicted by lies in the media'.

'People have trouble applying the notion of a good or bad consequence to *all the actual consequences that are good or bad*,' he said. 'Instead they see a small subset of consequences, the immediate local consequences, and think those are the "consequences".' For that reason, 'most people should not immediately try to be "utilitarians" ... They are better off continuing to debate which rules are good or bad and then following those rules.' For utilitarian reasons, don't try to be a utilitarian!

Again, it would amaze me if an internet guy in California had solved all the problems of moral philosophy. But I do find this approach refreshingly direct. There really is a moral law, of improving the world for the greatest number of people. It really does lead to some weird outcomes, like the torture/dust specks thing. However, it is a complex and difficult law to implement and we are usually best off implementing simpler, local laws, such as 'Do the thing that kills the fewest people.' You can contrive thought-experiment situations with trolleys or torture that end up forcing you into difficult situations, but in real life, 'Do the thing that kills the fewest people' is a solid position to take, and anything that steers you to a different answer should raise lots of red flags.

This is the basic moral position for the Rationalists: 'When human lives are at stake, we have a duty to maximise, not satisfice; and this duty has the same strength as the original duty to save lives. Whoever knowingly chooses to save one life, when they could have saved two – to say nothing of a thousand lives, or a world – they have damned themselves as thoroughly as any murderer.[10] And it has obvious implications for AI safety as well. Not simply that an AI that kills everyone is probably suboptimal from a utilitarian point of view, assuming that you agree that human lives are net-positive in the universe. There's also the discussion of what morals you instil in the AI itself: a 'friendly AI' that acts morally in the universe according to 'morals' that revolve around maximising happiness will be very different from one whose 'morals' revolve around maximising preferences, for instance. Also, an AI shorn of human biases might be more capable of implementing true utilitarianism, in a way that humans (and specifically me) apparently struggle with.

But most importantly, the Rationalist project is about encouraging 'rational' thinking – with, in Yudkowsky's case, the eventual goal of convincing everyone that there are good rational reasons to worry about AI safety. If you're going to think of the world in terms of numbers and statistics (which in Yudkowsky's view, and mine, is the only way you can make any sort of sensible decisions at a national or global scale), then you need a moral system that can give you numbers to plug in. Utilitarianism, with its harsh-seeming but impartial way of treating human lives as numbers, does that job neatly.

Part Four

Biases

Chapter 21

What is a 'bias'?

Part of Yudkowsky's project, in writing the Sequences, was explaining why AI might not look like human intelligence. Having had a go at explaining the basis of 'rationality' or 'intelligence' in its pure, general form, he then had to clarify why human intelligence wasn't quite the same thing. The most obvious reason is that humans are systematically biased, in ways that make us wrong in predictable directions. Over the next few chapters, we'll talk about a few things that make that happen.

We don't know everything about the world, and we never will. Not as individuals and not as a species. We just can't get hold of all the information.

But even when we *can* get hold of enough information about something to make a decision about it, we will sometimes be wrong in predictable ways, as a result of how the human mind works. The different ways in which our thinking goes wrong are often lumped together under the term 'cognitive biases'. Much of our understanding of them comes from the work of Daniel Kahneman and Amos Tversky, a pair of Israeli psychologists who did a series of groundbreaking experiments in the 1970s, although many other psychologists have worked on them since.

(A worthwhile caveat to mention at this point: since Kahneman and Tversky did their work, and since Kahneman's book *Thinking, Fast and Slow* made it especially famous in 2011, psychology in particular and science in general has been wracked by the 'replication crisis', in which many high-profile studies have turned out to be untrustworthy. Most of the stuff Kahneman and Tversky talked about is, I think, pretty robust, but it's just worth taking everything in psychology with a pinch of salt at this point.)

The Rationalists are extremely interested in all this. Yudkowsky started writing his Sequences on Robin Hanson's blog, which – you may recall – is called Overcoming Bias. The name LessWrong is a reference to avoiding, as far as possible, the biases which make us wrong. What we're not talking about is 'biases' in the 'football fan complaining that referees are biased against his team' or 'Donald Trump complaining about CNN' sense. We're talking about things that systematically reduce our accuracy in making guesses.

Rob Bensinger, in a foreword to one of the Sequences, gives an example of a (statistical) bias. Imagine, he says, that you have an urn with 100 balls in it – 70 white and 30 red – and you are allowed to take 10 of them out and then guess how many of the total are red or white. 'Perhaps three of the 10 balls will be red, and you'll correctly guess how many red balls total were in the urn,' he writes. 'Or perhaps you'll happen to grab four red balls, or some other number. Then you'll probably get the total number wrong. This random error is the cost of incomplete knowledge, and as errors go, it's not so bad. Your estimates won't be incorrect on average, and the more you learn, the smaller your error will tend to be.'[1]

But now, he says, imagine that the white balls are heavier than the red ones. They tend to sink to the bottom. 'Then your sample may be unrepresentative in a *consistent direction*', says Bensinger. Acquiring more data may not help you get it right. It may even make you more wrong.

Cognitive biases work in a comparable way. A cognitive bias 'is a systematic error in *how we think*, as opposed to a random error or one that's merely caused by our ignorance. Whereas statistical bias skews a sample so that it less closely resembles a larger population, cognitive biases skew our *beliefs* so that they less accurately represent the facts, and they skew our *decision-making* so that it less reliably achieves our goals.'[2] He gives the example of someone who has an optimism bias, and is told that the red balls can treat a disease that is killing that person's brother. 'You may then overestimate how many red balls the urn contains because you wish the balls were mostly red,' he writes. 'Here, your sample isn't what's biased. You're what's biased.'

There are various psychological reasons behind the individual biases, but the fundamental one appears to be that *they worked for our ancestors*. They were shortcuts. We didn't need to work out the value of 20,000 things compared to 2,000 things when we were tribal hunter-gatherers; we didn't need to work out probabilities. We could get pretty good estimates of values and risks from simple rules of thumb, or 'heuristics'. But now they often misfire.

Exactly what the biases in our minds are, and how they work, and which ones are separate from which, is an ongoing and probably unending project. But there are a few biases that most psychologists agree on, and – most importantly for this book – that are of interest to the Rationalists. I've picked out some examples, mainly from Yudkowsky's writing, as the sort of thing we're talking about. This is not an exhaustive list by any means, they're just the ones that I (subjectively) find the most interesting and important; although I did ask Yudkowsky if he agreed with my choices and he said, 'They sound like good guesses to me.'

The most important of all, though, is the last one we'll come to. If you remember any of them, remember that one.

Chapter 22

The availability heuristic

What's more likely to kill you: a terrorist attack, or the bath?

I'm not going to insult your intelligence. You know it's the bath, if for no other reason than the answer to 'What's more dangerous, [dangerous-sounding thing] or [not-dangerous-sounding thing]' is always '[not-dangerous-sounding thing]'. But I'm guessing most people, if asked to rank risks, would probably write 'terrorism' somewhere above 'bathtime'. After all, if you live in Britain, you'd have noticed no fewer than five high-profile terror attacks, four in London and one in Manchester, in 2017 alone. They probably wouldn't have heard of anyone dying in the bath.

But they'd be wrong and you'd be right. Over the last 10 years, there have been fewer than 50 deaths from terrorism on UK soil. (The large majority of them came in 2017, mostly in the awful attack on the Ariana Grande concert in Manchester.) That's an average of about five a year. According to an independent report on UK terrorism legislation carried out in 2012,[1] the average annualised death rate from drowning in the bath is 29.

This is an example of a systematic bias called the *availability heuristic*. When we are asked how likely something is, we could go and add up all the examples of it, divide this figure by the number of times it could possibly have happened, and get the answer. But that's difficult and takes a long time. What we *tend* to do, in reality, is to judge how likely something is by how easily we can think of an example; and how easily we can think of one is only loosely related to how often it happens. More dramatic things, which get disproportionate amounts of coverage in the media, are easier to remember. We can easily think of examples of terrorism, because every single one around the world gets reported, with dramatic images of smoke and fire and blood. We can't easily think of examples of drowning in the bath, because even though they happen far more frequently they don't make the news, and even when they do they're unspectacular.

Yudkowsky refers to a study[2] which looked at how good people are at assessing risks. It found that subjects 'thought that accidents caused about as many deaths as disease; thought that homicide was a more frequent cause of death than suicide. Actually, diseases cause about sixteen times as many deaths as accidents, and suicide is twice as frequent as homicide,'[3] he writes. This is a problem for various reasons. It leads to bad policies: if the public believes that child abduction is more common than it is, politicians will spend more money than they ought on reducing the risk; if people are more worried about Ebola than diabetes, then we might spend millions policing our airports to stop it coming in and neglect the thousands who die every year of diabetes. And it leads to bad personal decisions: in the years after 9/11, so many more people were afraid of flying, because of terrorism, that there were roughly 2,000 extra deaths on US roads; 300 a month in the first few months.[4] Less dramatically, we all know people who are afraid of visiting their city centres because of terrorist attacks, but don't think twice about driving to work.

This doesn't just apply to risk perception. You can easily think of examples of successful people because they're the ones in the news. 'In real life, you're unlikely to ever meet Bill Gates,' points out Yudkowsky. 'But thanks to selective reporting by the media, you may be tempted to compare your life success to his.' Your life is probably less successful than Bill Gates', by most measures, so that will make you sad. But, then, only one person in every 7 billion is Bill Gates. 'The objective frequency of Bill Gates is 0.00000000015, but you hear about him much more often. Conversely, 19 per cent of the planet lives on less than $1/day, and I doubt that one-fifth of the blog posts you read are written by them.'

The availability heuristic, like all other biases, presumably evolved because it was useful in the ancestral environment. A hunter-gatherer living in a tribe of 150 people would only have got news about those 150 people. You probably never heard of really unlikely things happening, because there weren't enough people for them to happen to. And dramatic, memorable things were probably worth remembering. As a yardstick for measuring objective probability, the availability heuristic most likely did a good job. But in a world of 7 billion people, instantly connected by the media, it can get things wildly wrong.

Of course, being aware of this doesn't stop it happening. I've known about the availability heuristic for years but I still look beneath me in the water when I'm snorkelling and imagine a shark coming up from the black depths. A hypothetical perfect Bayesian AI would assess the statistical likelihood of that and know that it is minuscule.

Chapter 23

The conjunction fallacy

What's more likely: that the climate will stop warming, or that a new technology will be developed which allows fuel to be economically harvested from atmospheric CO_2, and the ensuing reduction in greenhouse-gas levels stops the climate from warming?

If you're a relatively normal human being, you may find that option two sounds more likely. Option one feels a bit sparse. This is the normal reaction. The classic example that Yudkowsky cites[1] is a 1981 study by Tversky and Kahneman[2] which found that 72 per cent of subjects thought that 'Björn Borg will lose the first set' was less likely than 'Björn Borg will lose the first set but win the match', and 68 per cent of subjects thought that 'Reagan will provide federal support for unwed mothers and cut federal support to local governments' was more likely than that 'Reagan will provide federal support for unwed mothers'. By the way, it isn't the case (as I have always thought) that there were two groups, and one of them was asked for a probability on the first statement and another on the second. This was a group of people given a list of four possible outcomes, and ordering them from most to least probable.

You'll probably have noticed this already, but *it is impossible for them to be right*. It is impossible for 'Borg loses the first set but wins the match' to happen without 'Borg loses the first set' happening. It is impossible for Reagan to support unwed mothers, *and* cut support for local government, without supporting unwed mothers.

In mathematical notation, the probability P(A,B), that is to say the probability that both A and B will happen, *must* be lower than the probability P(B). If there's a 5 per cent chance that Borg loses the first set (Borg was very good, I gather), and an 80 per cent chance that, even having lost the first set, he still wins the match, then the chance of 'Borg loses the first set but wins the match' is 0.05 × 0.8 = 0.04, or 4 per cent.

The 'conjunction fallacy' is that adding details makes a story seem more *plausible*, even though they must – by the workings of mathematics – make it less *probable*. It happens to all of us, even professional forecasters. A separate study, also by Tversky and Kahneman, asked one group of analysts to rate the probability of 'A complete suspension of diplomatic relations between the USA and the Soviet Union, sometime in 1983', and another to rate that of 'A Russian invasion of Poland, and a complete suspension of diplomatic relations between the USA and the Soviet Union, sometime in 1983.'[3] The probability of the latter – which must, necessarily, be less likely – was judged to be higher.

We see the extra details as corroborative, says Yudkowsky (and Kahneman, and modern psychological science). But we should see them as *burdensome*. They don't make a story more likely, they make it less. People who want to avoid this 'need to notice the word "and"', says Yudkowsky. 'They would need to be wary of it – not just wary, but leap back from it . . . They would need to notice the conjunction of two entire details, and be shocked by the audacity of anyone asking them to endorse such an insanely complicated prediction. And they would need to penalise the probability substantially.' Again, humans don't do this; a perfect Bayesian AI would.

Chapter 24

The planning fallacy

How long will it take you to do something? Something big, some project that might require a few weeks or months?

A good rule of thumb: however long you think it will take, it'll probably take longer. (It might even be longer still. Douglas Hofstadter, the American polymath and author of *Gödel, Escher, Bach: An Eternal Golden Braid*, once coined 'Hofstadter's law': 'It always takes longer than you expect, even when you take into account Hofstadter's law.') That is because of a quirk of the mind known as the 'planning fallacy'. Yudkowsky mentions it in the Sequences.[1] He refers to a famous 1994 study by Roger Buehler and colleagues[2] which asked some students how long it would take them to complete their undergraduate theses. The students had to say when they were 50 per cent sure they'd finish their projects, 75 per cent sure, and 99 per cent sure.

'We found evidence of overconfidence,' Buehler writes laconically. Only 12.8 per cent of students finished their projects in the time they were 50 per cent sure they'd finish it by. Only 19.2 per cent finished it by their 75 per cent mark. And, amazingly, only 44.7 per cent even managed to get it done by their 99 per cent mark. 'The results for the 99 per cent probability level are especially striking,' says Buehler in the study. 'Even when they make a highly conservative forecast, a prediction that they feel virtually certain that they will fulfil, people's confidence far exceeds their accomplishments.'

Other scientists, including Tversky and Kahneman, have found similar results. What appears to be going on here is that if you ask someone how long something will take, they imagine all the steps that are involved and put a time on that. They don't include time for balls-ups or unforeseen disasters. Yudkowsky refers to another study,[3] which found that 'Asking subjects for their predictions based on realistic "best-guess" scenarios; and asking subjects for their hoped-for "best-case" scenarios produced *indistinguishable* results.' He continues: 'When people are asked for a "realistic" scenario, they envision everything going exactly as planned, with no unexpected delays or unforeseen catastrophes – the same vision as their "best case".' This may be the reason why, for instance, the retractable roof on the new stadium that the host city of Montreal built for the Olympics was not ready until 1989, 13 years after the Olympics had come and gone.[4] (And then it broke fairly shortly afterwards.)

There is a well-documented way *around* the planning fallacy, though. Don't just look at the specifics of what your project involves – look at how long other, similar projects have taken in the past. When I signed up to write this book I ascribed myself a six-week break in my full-time job, because I figured I could write the bulk of the 80,000 words in that period. Luckily I ended up going freelance (for journalists, this is usually a euphemism for 'getting fired', but that is *mostly* not true in my case) about six months before the deadline, and I used basically all of that time. If I'd spoken to a few of my peers who had written books before, I'd have noticed that they had all made similar assumptions, and ended up needing deadline extensions, and that actually writing a book takes for-bloody-ever. (I also found out, from one friend whose book came out fairly recently, that her publisher had accidentally CC'd her in on an email thread which said they'd given her a fake deadline, with the expectation that she'd *actually* submit her manuscript about three months later. Publishers, I suspect, know authors much better than authors do.)

This is called taking the 'outside view' instead of the 'inside view'. The 'inside view' is what you can see when you're looking at it from your own perspective. I know I can write a 2,000-word article in a day, so why can't I write 60,000 words in six weeks and do the rest at weekends or whatever? But the 'outside view' is what you find when you look at all the other people who've done similar things and see how long it's taken them. And books tend to take about a year to write.

Yudkowsky talks about this, as well. Buehler did another study,[5] which found (in Yudkowsky's write-up) that students 'expected to finish their essays 10 days before deadline. They actually finished one day before deadline. Asked when they had previously completed similar tasks, they responded, "one day before deadline".' 'So there is a fairly reliable way to fix the planning fallacy,' says Yudkowsky. 'Just ask how long similar projects have taken in the past, without considering any of the special properties of this project. Better yet, ask an experienced outsider how long similar projects have taken. You'll get back an answer that sounds hideously long, and clearly reflects no understanding of the special reasons why this particular task will take less time. This answer is true. Deal with it.'[6]

Chapter 25

Scope insensitivity

How much – in US dollars, or pounds sterling, or whatever – is a human life worth? And how much are a million human lives worth?

Whatever answer you give to the first question, the answer to the second – surely – should be a million times greater. That at least is the Rationalist response. For a lot of people, it may seem insensitive to talk about human life in monetary terms, but it has to be done, and in fact is done every day in the NHS and other healthcare systems. You need to know how much you can spend to save one life; otherwise, you'll spend far too much on one, and many others will die because you no longer have the cash to spend on them.

This may seem obvious, but in fact it is not. There is plenty of evidence to show that we are extremely inconsistent in our approaches to these things. 'Once upon a time,' writes Yudkowsky, 'three groups of subjects were asked how much they would pay to save 2,000 / 20,000 / 200,000 migrating birds from drowning in uncovered oil ponds.'[1] If the groups were approaching this *rationally* – if they attributed the same value to each bird's life – then whatever figure they gave for the first question, it should be 10 times as much for the second and 100 times as much for the third. This is not what they answered. They answered \$80 for the first question, \$78 for the second, and \$88 for the third.[2]

Yudkowsky points to similar experiments. Residents of four states in the western US said they would pay only 28 per cent more to protect all 57 wildernesses in the region than to protect just one. Toronto residents said they would pay about the same to clean up every polluted lake in Ontario as to clean up the polluted lakes in one part of Ontario. 'We are insensitive to scope even when human lives are at stake,' he says. 'Increasing the alleged risk of chlorinated drinking water from 0.004 to 2.43 annual deaths per 1,000 – a factor of 600 – increased willingness-to-pay [for measures to reduce the levels of chlorine in the water] from $3.78 to $15.23.'[3]

Exactly what's going on in our brains we don't know, obviously, but it appears that we make these judgements according to our emotional response, rather than any kind of numbers-based assessment. We picture a single, dejected bird, 'its feathers soaked in black oil, unable to escape', suggests Kahneman.[4] We imagine how that makes us feel, and put a dollar value on it. 'No human can visualise 2,000 birds at once, let alone 200,000,' says Yudkowsky, so we forget about that detail and just focus on the imaginary bird.

We also seem to care about the setting. You'd think a human life is worth a human life, and 5,000 human lives are worth 5,000 human lives, but instinctively we place them into a wider context. An intervention that would save 4,500 lives in a Rwandan refugee camp was considered far more valuable if the camp contained 11,000 people than if it contained 250,000, although the number of lives saved was the same.[5]

'There's a Jewish proverb,' Paul Crowley told me when I spoke to him. '"If you save a life, it is as if you've saved the whole world." And that's true. But then if you save two lives, it's as if you've saved two whole worlds.' He's borrowed this line from Yudkowsky, he tells me. We don't like thinking about this stuff. Even when we look at it logically, there's something icky about, for instance, saying that it costs too much to give *this particular child* an expensive experimental cancer treatment, so we have to let them die. But the Rationalists – with their shut-up-and-multiply, utilitarian-calculus ethic – are very good at thinking about it. And it's vital that we think about it, at least at a national level.

In the British NHS we have an organisation called the National Institute for Health and Care Excellence, or NICE. NICE's job is to determine whether or not the tax-funded NHS should offer treatments to patients. Periodically, we have a national uproar when some expensive new cancer drug is turned down, despite evidence that it works. A quick Google search found several such stories over the last few years, for example one published by my old employers the *Daily Telegraph*, which opens with the line: 'A "truly revolutionary" new drug that can give women with advanced breast cancer an extra six months of life will not be available on the NHS as it is too expensive.'[6] It's always easy, in those cases, to find some cancer patient who has been denied the drug and a chance of a longer life. But NICE works on a cost-effectiveness basis: it is willing to spend a limited amount, call it X, per quality-adjusted life year (QALY) saved. Spending 2X saving a QALY with a cancer drug means that they can't buy two QALYs' worth of diabetes drugs somewhere else. Being sensitive to scope means sometimes thinking, 'This person must die so I can save more people elsewhere', and that is never something we're comfortable talking about. But for Rationalists, who are temperamentally inclined to think in that way anyway, it is obviously vital to do so, if you're trying to achieve utilitarian goals in the world, be they reducing human suffering or avoiding human extinction.

Chapter 26

Motivated scepticism, motivated stopping and motivated continuation

Jonathan Haidt, the social psychologist, says in his (excellent) book *The Righteous Mind: Why Good People Are Divided by Politics and Religion* that when we are presented with evidence for or against a hypothesis, we ask ourselves one of two questions. When we want to believe something, 'we ask ourselves, "Can I believe it?" Then . . . we search for supporting evidence, and if we find even a single piece of pseudo-evidence, we can stop thinking. We now have permission to believe.'[1] But when we *don't* want to believe something, 'we ask ourselves, "Must I believe it?" Then we search for contrary evidence, and if we find a single reason to doubt the claim, we can dismiss it.'

So, says Haidt, when people 'are told that an intelligence test gives them a low score, they choose to read articles criticising (rather than supporting) the validity of IQ-testing. When people read a (fictitious) scientific study that reports a link between caffeine consumption and breast cancer, women who are heavy coffee drinkers find more flaws in the study than do men and less caffeinated women.' It even affects what you see: 'Subjects who thought that they'd get something good if a computer flashed up a letter rather than a number were more likely to see the ambiguous figure 🅱 as the letter *B*, rather than as the numbers 13.'

The technical terms for the 'can I believe it/must I believe it' phenomena are 'motivated credulity' and 'motivated scepticism'. Yudkowsky: 'A motivated sceptic asks if the evidence *compels* them to accept the conclusion; a motivated credulist asks if the evidence *allows* them to accept the conclusion.'[2] Yudkowsky adds another layer to this, which is the idea of *motivated stopping* and *motivated continuation.* When we're looking for something in real life, we aren't usually given a set of things to choose from: 'You have to gather evidence, which may be costly, and at some point decide that you have enough evidence to stop and choose. When you're buying a house, you don't get exactly 10 houses to choose from . . . You look at one house, and another, and compare them to each other [and] at some point you decide that you've seen enough houses, and choose.'

It's the same when you're trying to find the most likely hypothesis to explain some phenomenon, or the best answer to a question. Does a new drug reduce blood pressure? You can look at one study, but it might not be the whole story. You can look at another. How many should you read before you make a decision? But sometimes we have a reason to stop, or to continue, that isn't just about how much evidence is really necessary. However much evidence you have, you'll have a current best guess. You've looked at three studies and two of them say, cautiously, that the drug doesn't affect blood pressure; the third says, equally cautiously, that it does. Your current best guess might then be 'it doesn't work'.

But if you're a researcher at the company that makes the drug, you have a reason not to accept that conclusion, and to carry on looking for more evidence. '[When] we have a hidden motive for choosing the "best" current option, we have a hidden motive to stop, and choose, and reject consideration of any more options,' says Yudkowsky. 'When we have a hidden motive to reject the current best option, we have a hidden motive to suspend judgement pending additional evidence, to generate more options – to find something, anything, to do *instead* of coming to a conclusion.'

A real-life example that Yudkowsky quotes is that of the statistician R.A. Fisher, who argued (after the epidemiological evidence showed that smokers were vastly more likely to develop lung cancer) that smoking may not, necessarily, cause lung cancer. Instead he put forward as an alternative what became known as the 'genotype hypothesis', that people have a genetic tendency to want to smoke, and people with that genetic tendency are also prone to developing cancer.[3] Yudkowsky points out that Fisher may have had a 'hidden motive' to continue the search: that he was employed by tobacco firms as a scientific consultant. (For the sake of fairness to Fisher's memory, both biographies of which I am aware conclude that Fisher probably wasn't led by the money: 'This is to misjudge the man,' one states. 'He was not above accepting financial reward for his labours, but the reason for his interest was undoubtedly his dislike and mistrust of puritanical tendencies of all kinds; and perhaps also the personal solace he had always found in tobacco.'[4] That said, I suspect Yudkowsky would say, and I would agree, that this doesn't rule out the sort of subconscious bias that could affect his decision-making. And besides, while those motivations are not financial, they're not the disinterested seeking of truth either.)

Again, these aren't necessary features of all intelligence: they're specific flaws in *human* intelligence. A perfect Bayesian AI wouldn't have them; a more realistic, imperfect AI might have all sorts of flaws and idiosyncrasies in its thinking, but there's no reason to assume that they would be the same as ours.

Chapter 27

A few others, and the most important one

There are plenty of other biases, and if you want to find out more there are entire books dedicated to them. *Thinking, Fast and Slow* by Daniel Kahneman is a good one; Dan Ariely's *Predictably Irrational: The Hidden Forces That Shape Our Decisions* is another. Or, of course, you could sit down for several months with Yudkowsky's *Rationality: From AI to Zombies*, which I honestly recommend.

A few of the other biases that Yudkowsky mentions are the 'illusion of transparency',[1] in which we know the meaning of our own words, so we expect others to do so as well. For instance, in an experiment, subjects were told that someone went to a restaurant on the recommendation of a friend, and the restaurant turned out to be either a) horrible or b) nice. Then the diner left a message on their friend's answerphone, saying: 'I just finished dinner at the restaurant you recommended, and I must say, it was marvellous, just marvellous.'[2] Of the people who were told that the meal was horrible, 55 per cent said they thought that not only was the message sarcastic, but that the listener would *know* it was sarcastic. Of the people who were told the meal was nice, only 3 per cent thought it was sarcastic.

Relatedly, in 'hindsight bias'[3] people enormously overestimate how inevitable something was *after it happened*, or overestimate how obvious something is when they know the answer. People, for instance, frequently assume that social-scientific results are statements of the obvious – but experiments have shown that they would say that *whatever the result actually was*. For example, if you give half of the subjects in an experiment the following quote:

Social psychologists have found that, whether choosing friends or falling in love, we are most attracted to people whose traits are different from our own. There seems to be wisdom in the old saying 'Opposites attract'.

and the other half:

Social psychologists have found that, whether choosing friends or falling in love, we are most attracted to people whose traits are similar to our own. There seems to be wisdom in the old saying 'Birds of a feather flock together.'

then 'virtually all will find whichever result they were given "not surprising".[4]

'Loss aversion' is where we assign more value to things we *have* than things we can *get*, so we might refuse to bet £1 on a flipped (fair) coin with the chance of winning £3.

The 'affect heuristic' is our tendency to assume that if something is good in *one* way, it's good in *all* ways. 'Subjects told about the benefits of nuclear power are likely to rate it as having fewer risks,' writes Yudkowsky. 'Stock analysts rating unfamiliar stocks judge them as generally good or generally bad – low risk and high returns, or high risk and low returns – in defiance of ordinary economic theory, which says that risk and return should correlate positively.'[5] The 'halo effect' is when the affect heuristic is applied socially: so if someone is handsome, we tend to assume that they're also intelligent and moral.[6]

But the most important bias to be aware of is this, which is a sort of collection of several: *knowing about biases can make you more biased.* Yudkowsky:

Once upon a time I tried to tell my mother about the problem of expert calibration, saying: 'So when an expert says they're 99 per cent confident, it only happens about 70 per cent of the time.' Then there was a pause as, suddenly, I realised I was talking to my mother, and I hastily added: 'Of course, you've got to make sure to apply that scepticism even-handedly, including to yourself, rather than just using it to argue against anything you disagree with—'

> And my mother said: 'Are you kidding? This is great! I'm going to use it all the time!'[7]

Various biases can actually mean that even as you get more information, you become more wrong. Confirmation bias and disconfirmation bias, and related phenomena, for instance. New information comes in, but your brilliant mind finds brilliant ways in which to ignore the stuff it doesn't like and promote the stuff it does.

There's a particularly pernicious one, the 'sophistication effect': 'Politically knowledgeable subjects, because they possess greater ammunition with which to counter-argue incongruent facts and arguments, will be more prone to [these] biases.'[8] So new information like 'We are all biased and the things we believe are frequently wrong' can easily become 'These arguments that are being deployed against me are flawed, and I can point out why because I have this in-depth knowledge of human biases.' Yudkowsky calls this a 'fully general counter-argument'. Anybody with a partisan axe to grind can deploy 'confirmation bias' to undermine an argument they don't like. Most of the things we call 'human biases' are extremely convenient labels to attach to opinions with which we disagree.

But the key is to accuse *your own* of them. *You* are biased. (*I* am biased.) *You* are probably systematically overconfident in your beliefs. I certainly am: in fact I took a calibration test on the Good Judgment Project website recently which showed that, yes, I overestimated my knowledge of economics, geography, history and world politics. (I was well calibrated for general knowledge and underconfident in my knowledge of Europe, if you're interested.) This isn't stuff that you should just be applying to *other people*. You need to apply it to *you*.

Paul Crowley joked about this with me, when I spoke to him. When he first started reading Yudkowsky's Sequences, he said, 'I read it and said, "This is brilliant! It shows how everyone apart from me is wrong." And then you read a bit more, and you think, *hmm, you know, maybe some of this might apply to me.* The mote in the other person's eye is easier to see.'

The Yudkowsky/LessWrong/Rationalist project is to help people to see those motes in their own eyes, in order to help them behave more like perfectly rational Bayesian optimisers.

Part Five

Raising the Sanity Waterline

Chapter 28

Thinking probabilistically

A large part of the Rationalist project is how to *improve your own rationality*. That is, how to get closer to making Bayesian-optimal decisions and holding true beliefs, given the constraints of the human brain and its many biases. Yudkowsky dedicates large parts of the Sequences to this 'martial art of rationality'.

As we saw in the sections on Bayesianism and utilitarianism, the Rationalist movement likes to put numbers on things, even if those numbers are estimates. A significant part of that is putting explicit figures on how *likely* you think something is.

In 2015, a book called *Superforecasting: The Art and Science of Prediction* came out. It was about telling the future, and my Rationalist friend Paul Crowley was very excited about it. 'I felt like that was a vindication of everything we'd been talking about for 10 years,' he said. *Superforecasting* was by Philip Tetlock and Dan Gardner, and was a write-up of Tetlock's work as a professor of political psychology at the University of Pennsylvania.

In 1984 Tetlock, a recently tenured professor, was asked to work on a new committee appointed by the National Academy of Sciences. Its goal was to help stop nuclear war. Tensions were enormously high between the two superpowers; Stanislav Petrov, whom you may remember from the section on existential risks, had (although no one on the committee would have known) quite possibly saved the world just a few months before. Tetlock sat on the committee with other well-respected social scientists, who argued over how best to reduce the risk of confrontation. He told Gardner, years later: 'I mostly sat at the table and listened ... The liberals and conservatives in particular had very different assessments of the Soviet Union. The conservative view was that they could . . . contain and deter it. Whereas the liberal view [was] that conservatives [in the White House] were increasing the influence of the hardliners in the Kremlin.'[1]

A few months later, Mikhail Gorbachev took command of the USSR, and started implementing liberal policies. No one had expected it, but both liberals and conservatives took it as confirmation that they had been right all along. 'The conservatives argued that we had forced the Soviets' hand,' said Tetlock. 'Whereas [the liberals thought] the Soviet elite had learned from the failings of the economy [and that] if anything, we had slowed down the process of learning and change.'

Tetlock came to the conclusion that what *actually happened* had very little bearing on whether or not the experts judged their predictions to be right. They just explained what happened in terms that made the stories they were telling seem true anyway. He was intrigued by this, so he set up an experiment. He recruited hundreds of experts from various fields – journalists, economists, political scientists – and asked them for their anonymous predictions.

Part of the problem, Tetlock had noticed, was that people often gave predictions with ambiguous interpretations, about 'growing tensions being likely' or suchlike, that didn't really tie them to any specific outcome. So the questions he asked had easily confirmed answers and clear timeframes: 'Will the dollar be higher, lower, or the same against the pound a month from now?' 'Will North Korea and the United States go to war in the next two years?' And the experts had to give precise numerical estimates of how likely they thought the outcomes were: 30 per cent chance, 75 per cent chance, 99 per cent chance, and so on.

He collected nearly 30,000 predictions from 284 experts. He waited, weeks, months and in some cases years, to see how well they did against the harsh judgement of reality. And he counted how often their predictions matched reality. If someone's 75 per cent predictions came up 75 per cent of the time, and their 90 per cent predictions came up 90 per cent of the time, and so on, then they were 'well-calibrated'.

He also gave them a bonus score for being precise. If you just said

'50 per cent chance' for everything, then you'd probably do OK at calibration. But you'd be no use as a predictor; we want someone to state, 'This will happen' or 'This won't happen'. So saying that something is 99 per cent likely gets you a higher score, if you're right, than saying that something is 60. That's called 'discrimination'.

The results, when his study was published years later, are pretty famous: on average, the experts were no better calibrated than, as Tetlock put it, 'a dart-throwing chimpanzee' – literal random chance. (They were a bit better at discrimination, but not really.) What was interesting, though, was what happened when you divided them up further. Some experts not only did as badly as the imaginary chimpanzee, they were significantly worse; they really would have improved their scores if they'd answered at random. But some did much better. 'There's quite a range,' Tetlock told Gardner. 'Some experts are so out of touch with reality they're borderline delusional. Other experts are only slightly out of touch. And a few experts are surprisingly nuanced and well-calibrated.'[2]

It wasn't their political views that best predicted who did well, or even their level of education or experience in the field. Instead, it was the way they thought. The ones who did amazingly badly were those who believed that there was a big idea which explained everything; that the world was simple and could be understood simply, that they could just stamp their big idea onto every situation. The ones who did well were those who had no such big idea, who regarded the world as complex, took their information from many different sources and were willing to be self-critical and learn from mistakes. Tetlock called the former 'hedgehogs' and the latter 'foxes', following an Isaiah Berlin essay quoting an old Greek poem: 'The fox knows many things, but the hedgehog knows one big thing.'[3]

Years later, Tetlock co-operated with the US Defense Department's Intelligence Advanced Research Projects Activity (IARPA) to run a competition to find the best forecasters; they outperformed actual CIA operatives by a margin, even with no access to classified information. The Good Judgment Project spawned the *Superforecasting* book; the very best forecasters were known as superforecasters.

The reason why Paul Crowley was excited about it was that this is exactly the sort of thinking that Yudkowsky and the Rationalists have been talking about for ages. It's pure Bayesianism. 'Look, it turns out thinking probabilistically is really important for making accurate guesses about things!' he said, happily. 'These guys are explicitly being Bayesian. They're using priors, they're updating with evidence, they update in a Bayesian way.'

Say you've been asked to predict whether there'll be a war between North Korea and the US in the next year. You might look at the war of words – as I write it's a few months since President Donald Trump called Kim Jong-un 'rocket man' and accused him of being 'short and fat', for instance, since that's the Churchillian rhetoric of the era we live in – and conclude that, well, it feels pretty likely. Then you might put a figure on that feeling, and say '30 per cent'. But a Bayesian – a superforecaster – would try to find prior probabilities, and look at other forms of evidence.

You might try to find a prior probability for a war with North Korea by, for example, researching the number of wars between the two since the Second World War. That's one war in 70 years, so your prior probability for a war in any given year is low – about 1.5 per cent. That's the equivalent of the 'background rate of cancer' that we saw in the Bayesian explainer a few chapters ago. Then you could look at all the times that Trump has tweeted aggressively at world leaders. Say he goes to war within a year with world leaders at whom he tweets aggressively 90 per cent of the time; he only goes to war with those at whom he tweets non-aggressively 15 per cent of the time. But you also know that wars between North Korea and the US are pretty rare. In every 100 years, you get about 1.5 wars. So your rate of new wars per year is about 0.015.

You can plug all these numbers into the equation, exactly the same as the cancer test. Your war has an incidence rate of 0.015. So for every million world leaders Trump tweets at, he will declare war on 15,000 of them within a year. The test ('Has Trump tweeted aggressively?') will accurately pick out 13,500 of them. And of the 985,000 world leaders at whom Trump tweeted more soothingly, 837,250 will be told, correctly, that they are not going to be targeted by cruise missiles. But 147,750 will be told – wrongly – that they *are* going to have a war. That's your false positive rate. So you have a total of 161,250 positive results (belligerent tweets), of which 13,500 are true and 147,750 are false. So your odds of a war with North Korea, given that Trump has tweeted belligerently at Kim Jong-un, are a bit over 8 per cent. (Also, 1,500 world leaders are going to have a heck of a surprise after Trump tweets 'Great guy!1!' at them and then launches a series of Harpoon missiles at their oil refineries.)

In real life, you wouldn't have good numbers like this. Your prior of wars per year might be pretty solid, but you don't have a huge database of Trump tweets and Trump wars (not least because, however much most people reading this book probably dislike Donald Trump, at the time of writing he hasn't started many wars). But Tetlockian 'foxes' would use best-guess numbers to fit the various bits of the Bayesian equation; they would look for other sources of information to adjust their numbers up and down; they would allow their estimates of the probability to move where the evidence then took them. They might not explicitly run through Bayes' theorem in their minds, but they would do something analogous.

And the key thing is then *checking whether you are right.* There's a difficulty, of course. If you predict that there's an 8 per cent chance of war, that doesn't mean that you're saying there won't be a war; you're saying that it's unlikely, but that there's still about a one in 12 chance. So if there is a war, you could reasonably claim, 'Well I didn't say there wouldn't be,' and mark yourself as correct.

The way around this is to make lots of predictions and see how many come in. This method became particularly famous around the 2012 US presidential election, when Nate Silver, editor-in-chief of the website FiveThirtyEight.com, correctly predicted which way all 50 states would end up voting. He did it using exactly these methods: having a prior and then updating it with evidence, in the form of new polls. (The new polls were added to the database using an algorithm, thus taking human bias out of the equation, to a degree.) The site would – and still does – give percentage estimates for each event it predicted. The idea is that it can go back and grade itself: as we saw earlier in Tetlock's experiment, if its 75 per cent bets come in 75 per cent of the time, then it's well calibrated.

And this is explicitly what the Rationalist community does. Scott Alexander of Slate Star Codex does it every year: he makes predictions in January, and the following January he grades them to see how well calibrated he is. For instance, at the start of 2017 he predicted with 60 per cent confidence that the US would not get involved in any new major war with a death toll of more than 100 US soldiers, with 95 per cent confidence that North Korea's government 'will survive the year without large civil war/revolt', and with 90 per cent confidence that 'no terrorist attack in the USA will kill more than 100 people'. He made 104 predictions; at the start of 2018 he went back and looked at them, and checked his calibration. (He did pretty well. His 60 per cent predictions came in 64 per cent of the time, his 70 per cent predictions 62 per cent of the time, and so on.)

And as a whole, the Rationalists are really good at this stuff: they tend to be foxes rather than hedgehogs. I spoke to Michael Story, a superforecaster who works for Tetlock's Good Judgment Project. I asked him exactly what that meant. There are 20,000 forecasters in their sample, he said. 'That's how many forecasters, of whom 150 are supers,' he told me. And are you one of the supers, I asked? 'I am indeed,' he said, somewhat shyly. (I met Mike in a café in north London for breakfast. He's tall and bearded and extremely friendly, but my favourite thing about him is that he has this ridiculously enormous dog called Laska. Laska is an Alaskan shepherd; he looks like a wolf and weighs more than 10 stone. He sat on my foot for a bit – Laska, not Mike – and my foot went to sleep.)

You can see from their analytics where people come to the Good Judgment Project from, Mike told me, whether they click through from the *Guardian* website or Google or whatever. 'Loads of the superforecasters came from LessWrong,' he said. 'A ridiculously disproportionate number. Same with a lot of them, if you trace back how they first got involved, loads of them will say LessWrong or blogs associated with it, [Tyler Cowen's] Marginal Revolution or [Robin Hanson's] Overcoming Bias, that crowd.' His impression

– and he's careful to say that it's just his impression – is that superforecasters and Rationalists are similar in a lot of ways. 'My impression is that supers and LessWrongers share similar norms of open discussion, and probably similar personality types, especially with attitudes to conflicting arguments and information. I only have data on superforecasters, so I can't compare directly, but I've noticed many of the same themes emerging.'

Mike is a long-term fan of the Rationalist community ('I unironically love it'). He's met many of the people mentioned in these pages, and been to their IRL (that's 'in-real-life', for those of you who don't live on the internet) parties in Oxford and elsewhere. And he thinks that the Rationalists are so good at forecasting because the community has the norms it has – of free speech, and accepting weird and outré and even offensive views. It is, after all, harder to hold on to any one big idea about why things happen when you are surrounded by people who think your big idea is stupid.

Obviously, they have their own big idea that the world may well be destroyed by AI. There are two possible responses to this. One is to point out, correctly, that this is a thing they forecast, rather than something they plug into the forecast; it's not something that they can stamp on every situation and then say, 'I think France will lower its top rate of tax in the next 18 months because the world is going to be destroyed by AI.' You could say that the fact that a lot of them are superforecasters is itself evidence in favour of the hypothesis 'The world may be destroyed by AI.'

The other response is to say, as Mike does, that it could be their weakness. For all that they allow very different political beliefs into their sphere, the Rationalists are, as a rule, very similar. They're non-diverse in the ways that get people angry – they're predominantly white and male – but they're also similar in personality types: nerdy, often autistic, scoring high on personality traits such as scrupulosity and conscientiousness, often introverted. And that, said Mike, could be 'dangerous'. 'This is my concern. If everyone's too similar, you're vulnerable to a bad meme, just the same as biologically if you have all these plants that are the same, one virus kills them all.' I asked him if he thought that the AI stuff was a 'bad meme' that has got into the Rationalist ecosystem and now can't be eradicated because everyone is too similar, and he said that he wasn't sure. But it is worth worrying about, he said. 'If everyone's personalities line up, like holes in Swiss cheese, then everyone could adopt a bad meme and not realise.'

Chapter 29

Making beliefs pay rent

Another key way of checking your own beliefs is to think about what they actually imply. Yudkowsky calls this 'Making beliefs pay rent in anticipated experiences.' For instance: if a tree falls in the forest, does it make a sound? Answer that question in your mind before you go any further. If you thought 'no', is that because, to you, 'sound' means the sensation, the qualia, of someone hearing something? And if you thought 'yes', is that because 'sound' means the pressure waves in air that are made when something loud happens?

This is one of the longest-running arguments in philosophical history, to the point that it's a cliché of philosophy alongside angels dancing on the head of a pin. But, assuming that you agree that the physical world *still exists* when we are not looking at it (which some philosophers dispute, but I am content to ignore them), then – what are people *actually arguing about?*

Yudkowsky imagines an argument between two people, Albert and Barry:

> **Albert**: 'What do you mean, there's no sound? The tree's roots snap, the trunk comes crashing down and hits the ground. This generates vibrations that travel through the ground and the air. That's where the energy of the fall goes, into heat and sound. Are you saying that if people leave the forest, the tree violates Conservation of Energy?'
> **Barry**: 'But no one hears anything. If there are no humans in the forest, or, for the sake of argument, anything else with a complex nervous system capable of "hearing", then no one hears a sound.'[1]

But, points out Yudkowsky – who imagines the argument spiralling out of control somewhat – Albert and Barry actually agree on everything that is happening. They both think that the tree hits the ground and sends waves of energy through the forest. They both agree that no auditory sensations are being experienced. All they disagree about is whether or not that combination of things should be called a 'sound' or not. If you had two words – Yudkowsky suggests 'albergle' for acoustic vibrations, 'bargulum' for auditory experiences – then the argument would disappear; they'd just say 'OK, it makes an albergle but not a bargulum'.

A surprising number of arguments seem to fall into this form. (About 40 per cent of those on the contemporary British internet seem to revolve around whether or not Person A or Group B is Marxist/socialist/Nazi/alt-right/misogynistic/racist/transphobic/a TERF etc., with people on each side marshalling reasons for and against their inclusion in one definition or another.) But these debates are sterile, for Yudkowsky and the Rationalists, because they don't *constrain your expectations*. If your model can explain *every* outcome, then it can't explain any outcome.

If I argue that we should define 'sound' as 'acoustic vibrations' rather than 'auditory experiences', it won't change what I expect to find when I walk into the forest to see where the tree has fallen. If I argue that we should define Jeremy Corbyn as a 'Marxist' rather than a 'socialist', it won't change what I expect him to do if his Labour Party is elected to power. If I say, 'I believe that the tree's trunk broke, rather than that the roots came out of the ground', that is a belief that constrains my experiences; if I turn up and see that the roots are out, then I know that my belief was wrong. 'I believe that Jeremy Corbyn will renationalise the British railway system within a year of coming to power' constrains my experiences; if he does not, then I know I was wrong. But 'Jeremy Corbyn is a Marxist' does not constrain my beliefs and cannot be used to predict anything: if Corbyn does *not* nationalise the railways, he could still be a Marxist, and vice versa. ('But we would expect more Marxist-style behaviour such as compulsory nationalisation from someone who is a Marxist!' Fine, but in that case what you *call* him doesn't matter. What behaviour do you expect?)

'When you argue a seemingly factual question, always keep in mind which difference of anticipation you are arguing about,' says Yudkowsky. 'If you can't find the difference of anticipation, you're probably arguing about labels. Above all, don't ask what to believe – ask what to anticipate. Every question of belief should flow from a question of anticipation, and that question of anticipation should be the centre of the inquiry. Every guess of belief should begin by flowing to a specific guess of anticipation, and should continue to pay rent in future anticipations. If a belief turns deadbeat, evict it.'[2]

Chapter 30

Noticing confusion

There's an old science joke which Yudkowsky turns into a Teachable Moment. There is a heater in the laboratory. Next to it is a tile. The teacher asks her students: 'Why do you think the side of the tile *next* to the heater is cooler than the side *away* from the heater?' (If you like, stop reading for a moment and think why it might be. Don't feel you have to, though.)

The student stammers: 'Er, perhaps because of heat conduction?'

And the teacher replies: 'No, it's because I turned the tile around before you came in.'

It is a god-awful joke, I realise, but it is useful. The phrase 'because of heat conduction' *sounds like* an explanation, says Yudkowsky. It fits into that bit of the conversation where an explanation would go, and it uses sciencey-sounding words. But remember the last section, about making beliefs pay rent in anticipated experiences. What does a belief in 'heat conduction' make you *expect*?

Well, it should boil down to a series of equations derived from Fourier's law and the conservation of energy (I say, confidently, having checked Wikipedia). But to a first approximation, it should say that the bit that's been heated up should be hottest, and that bits that are further away should be cooler. The student should expect to find that the side nearer the heater is warmer, and should be surprised when it isn't. 'If "because of heat conduction" can also explain the radiator-adjacent side feeling *cooler*,' says Yudkowsky, 'then it can explain pretty much *anything*.'[1] And if your model can explain everything, then it doesn't explain anything. When something happens that your beliefs don't anticipate, you should be *confused*. And you should pay attention to your confusion, because either your belief model is wrong, or something else is going on that you're not aware of, like the tile being turned around.

Yudkowsky has a story of someone telling him in a chatroom that a friend needed medical advice: 'His friend says that he's been having sudden chest pains, so he called an ambulance, and the ambulance showed up, but the paramedics told him it was nothing, and left, and now the chest pains are getting worse. What should his friend do?' Yudkowsky says he knew that paramedics don't do that – that if someone calls an ambulance they are obligated to take them to the emergency room – but didn't take the obvious next step. Instead, he managed to 'explain the story within my existing model, though the fit felt a little forced', and replied: 'Well, if the paramedics told your friend it was nothing, it must really be nothing – they'd have hauled him off if there was the tiniest chance of serious trouble.' Then it turned out that the friend had made the whole thing up.

'My feeling of confusion was a clue,' says Yudkowsky, 'and I threw my clue away. I should have paid more attention to that sensation of *still feels a little forced*. It's one of the most important feelings a truth can have, a part of your strength as a Rationalist. It is a design flaw in human cognition that this sensation manifests as a quiet strain in the back of your mind, instead of a wailing alarm siren and a glowing neon sign reading: Either Your Model Is False Or This Story Is Wrong.' If you're trying to become a more rational being – a better Rationalist – then you need to listen to those little moments when something doesn't quite seem to add up.

Chapter 31

The importance of saying 'Oops'

The key takeaway from all the 'bias' stuff we've talked about is probably that *it is really hard to change your mind*. Large parts of our make-up are geared towards letting us keep on thinking what we already thought: confirmation bias, motivated reasoning, loss aversion and so on. If someone tells you something you don't want to hear, then you'll find ways of not believing them. That's why a major part of the LessWrong project is learning *how to actually change your mind*. A whole Sequence of the Sequences, in fact, bears that exact title.

'Scott [Alexander] said something I thought was really central to our enterprise,' Paul Crowley told me. 'Just as it's good to have a lot of money, it's good to have as much evidence as possible. But just as it's good to get by on how much money you have, it's good to be able to be as accurate as possible with the evidence you have. Sometimes the universe is not going to lavish you with evidence, sometimes you have to be as accurate as you can, with what you can get.' And that means getting rid of ideas when the balance of evidence is against them, rather than – as our biases would have us do – hanging on to them for as long as we can. 'We wait until we're overwhelmed,' said Paul. 'It's a long, slow process.' But instead, we ought to treat *ideas we hold* and *ideas we do not hold* equally with respect to the evidence.

Quite a few of the Rationalists seem to come from religious backgrounds, and a key moment in their Rationalist life story is the point at which they gave up on religion, as they realised the evidence did not support it. 'I haven't had that one,' Paul told me, laughing. 'I was brought up by atheists. But I've had something similar.' His parents are both socialists, and he followed them.

'I was a card-carrying revolutionary communist in 1989, 1990,' he said. 'Then I moved towards being a more classically wishy-washy socialist type, wanting to achieve socialism by democratic whatever, but largely trying to come up with a position I could defend.' Then, while reading the Sequences, he came upon the post entitled: 'The importance of saying "Oops"'.[1]

He'd been edging away from his socialist beliefs, retreating, 'fighting a rearguard action', as he put it. 'But I felt like, on reading that, there was a level on which I already knew that this didn't make sense, that I couldn't sell it any more. I read it and just went, like, no. When you're fighting the rearguard action, stop fighting. Stop, and reassess. Sometimes you have to say, "I've made a large mistake."' Instead of seeking the closest defensible position to your current one, try to ask where the evidence points, and sit in the middle of that.

When Yudkowsky slowly moved from his original 'the singularity will solve everything' position to his later 'AI might actually destroy everything', he did it incrementally. 'After I had finally and fully admitted my mistake, I looked back upon the path that had led me to my Awful Realisation,' he wrote. 'And I saw that I had made a series of small concessions, minimal concessions, grudgingly conceding each millimetre of ground, realising as little as possible of my mistake on each occasion, admitting failure only in small tolerable nibbles. I could have moved so much faster, I realised, if I had simply screamed "Oops!"'[2]

None of these techniques or tricks are ever going to turn a kludgy human intelligence into the pure Bayesian thought-being. But Yudkowsky's hope (and the wider Rationalist project) is that by using them, people will be better able to assess ideas and decisions, on both a personal level and a societal one. (One of those decisions, he thinks, will be to take the issue of AI alignment seriously.)

Part Six

___

Decline And Diaspora

Chapter 32

The semi-death of LessWrong

The Yudkowsky project we've been discussing, of explaining rationality, human thought, where they differ and how to make the latter more like the former, grew out of SL4 and Overcoming Bias, and became the Sequences and LessWrong, and the Rationalist community.

LessWrong was its central hub for a long time. But in about 2012 LessWrong started to die off – not completely, but its numbers dropped significantly. A peak of a million or so page views a month in early 2012 dropped to about 350,000 a month by mid-2016.[1] There are lots of reasons behind the decline, but here are the main two. One, Eliezer Yudkowsky felt that he'd finished the 'Sequences' at some point a couple of years earlier, towards the end of 2010. So he stopped blogging. And two, in 2013 Scott Alexander – known on LessWrong as Yvain, and probably the most prominent blogger on the site after Yudkowsky himself – started his own blog, Slate Star Codex.

There were other reasons. Robin Hanson told me that he thinks it's partly because, as with many things, the Rationalists tried to reinvent everything from scratch. LessWrong, the website, wasn't just a blog: it was a custom-designed community hub, based on a Reddit-like voting system – if you like a post or a comment, you press the 'up' arrow; if you don't, you press the 'down'; the site's algorithm is more likely to show you things with lots of upvotes than things with lots of downvotes. But it got gamed: according to Scott Alexander, 'one proto-alt-right guy named Eugene Nier found ways to exploit the karma system to mess with anyone who didn't like the alt-right (i.e. 98 per cent of the community) and the moderation system wasn't good enough to let anyone do anything about it.'[2]

It's part of a wider attitude among the Rationalists, said Hanson, of thinking they can rebuild everything. 'They're people who are smart and articulate and they have ideas in their heads about how things should be different. And they want to implement them all, and are unpersuaded by the fact that other people have tried them before and failed.' The Rationalists, unsurprisingly, do not agree with this assessment, but there is an element of truth to it.

So they convinced themselves, Hanson thinks, that by studying the Art of Rationality – the 'Bayesian judo' that Yudkowsky was teaching, all that stuff about noticing confusion and thinking probabilistically – they could avoid the pitfalls of irrationality that flaw other people's thinking and create new, shiny things. 'They decided that they could make better technology,' he said, 'like the LessWrong software. They get involved in start-ups, they think they know how to redo romance with polyamory, they think they know how to redo diets with the diet things they get into. They're all over the place with whatever.' It reminded him, he said, of some Silicon Valley people he was involved with in the 1980s, the Xanadu Project, who were working on the first hypertext systems, and some others who were interested in nanotechnology. 'There were a lot of young idealistic people trying to save the world through start-ups and tech. As usual they were into science fiction and the future and how everything would change enormously, and they were into trying everything different. At Xanadu they had to do everything different: they had to organise their meetings differently and orient their screens differently and hire a different kind of manager, everything had to be different because they were creative types and full of themselves. And that's the kind of people who started the Rationalists.'

And Scott Alexander has his own theories, which he expounded on a Reddit thread in 2017. One was that Yudkowsky skirted 'the line between "so mainstream as to be boring" and "so wacky as to be an obvious crackpot"', which many other bloggers struggled to do, either because they were boring, or because they were crackpots, or because although they weren't crackpots they also weren't very good at not coming across as crackpots. Yudkowsky also came to realise, Scott said, that he is 'a pretty weird person, and now that the community's more mature it helps for it to have less weird figureheads'.

Scott added that the community became 'an awkward combination of Google engineers with physics PhDs and three start-ups on one hand, and confused 140-IQ autistic 14-year-olds who didn't fit in at school and decided that this was Their Tribe Now on the other', and that it was hard to find the 'lowest common denominator' that appealed to both groups. The end result was that 'LessWrong got a reputation within the Rationalist community as a bad place to post, and all of the cool people got their own blogs, or went to Tumblr, or went to Facebook, or did a whole bunch of things that relied on illegible local knowledge [by which he means the sort of understanding of a community that only comes from living in it – like how you'll always know your home city better than someone who's just read the *Lonely Planet* guide to it]. Meanwhile, LW itself was still a big glowing beacon for clueless newbies. So we ended up with an accidental norm that only clueless newbies posted on LW, which just reinforced the "stay off LW" vibe.'

The point about 'illegible local knowledge' is definitely true from my experience. I've been hanging around LessWrong-ish circles for a few years now, and I regularly still stumble across giants of the Rationalsphere whom everyone else just seems to know but who never wandered into my field of vision before: names like Gwern and Nostalgebraist and The Unit of Caring. There's no natural central hub any more, and you can't learn the paths and backwoods of Rationalist country without wandering around it, lost, for years. (There is a map. It's by Scott Alexander. But I don't think you could use it to navigate without first knowing where everything is anyway. Still, the puns – 'Reasoning Sound', 'Bight of Information', 'The Reverend Thomas Bay' – are absolutely impeccable.)[3]

Whatever caused the semi-death of LessWrong, the fact remains that it happened, and the people involved in it spread across different parts of the internet. Slate Star Codex gets about 20,000 views a day, about 600,000 per month – down from the LessWrong peak, but still a significant number.

Chapter 33

The IRL community

What also happened is that a real-life community sprang up. There are Rationalist (and 'Rationalish', 'Rationalist-adjacent', LessWrong and Slate Star Codex) meet-ups all around the world: I can see Facebook groups for Munich, the Netherlands, Israel, Montreal, London, Reading, Bath, Sydney, Denver, DC, Canberra, Edinburgh, Darmstadt and Phoenix, just via a very quick search. I know there's a group in Melbourne as well, and Berlin, and a nascent one in Manchester. The Bay Area and New York have the largest communities.

Many of the community members live in group houses, although not all or even a majority. This doesn't appear to have been a cult-leader invention enforced from the top by Eliezer Yudkowsky; Paul Crowley thinks that 'the group house phenomenon is just a necessity given the Bay housing situation' (house prices there being, to my amazement, comparably extortionate to London), which is where a large number of them live, and that, anyway, Rationalists are the sort of people who tend to end up in group houses. 'The libertarian/futurist/Burner [Burning Man festival regular] circles were in named group houses long before Eliezer moved here,' he said. 'At university in the 1990s my social circle was the science-fiction society, and we all lived in named group houses.'

Ben Harrison, a British man in his early twenties who got involved with the Rationalists through reading the blogger Gwern and ended up following the LessWrong/Slate Star Codex stuff, told me that the group houses are 'a bit like university halls, but the kitchen sink is a little cleaner'. His group house, in Manchester, was set up explicitly as an alternative to the Bay Area, which tends to attract many Rationalists, but because of the aforementioned housing costs has a high bar to entry. Some of the group houses are polyamorous; some are not.

I wanted to get an idea of what the IRL community was like, so I went to a few of the meet-ups. I met Scott Alexander, the AI researcher Katja Grace, who was in a polyamorous relationship with Scott at the time, and a couple of others at a pizza place in Berkeley while I was there, where we discussed AI safety and whether or not my book was going to be a catastrophe that increased the likelihood of a paperclip apocalypse. (Buck Shlegeris, a young MIRI employee with excitingly coloured hair and an Australian accent, told me that 'A book on this topic could be good', and that 'if I could jump into your body I have high confidence I could write it'. However, his confidence that *I* could write it from within my *own* body seemed significantly lower, which is probably fair enough.)

I distinctly got the impression that the IRL community is, like the online community, a venue for people who are a bit weird, not very good at small talk, and interested in big ideas. There were a couple of things that stood out for me while I was there. One was that, for a few minutes, I couldn't find Katja. Then it turned out that was because she was sitting with a baby on her lap. I knew she didn't have a child, so I'd discounted the woman with the baby as obviously not her. Upon closer inspection, it turned out that the baby was one of those robot babies that some American high schools give out to teenagers, to give them an understanding of how hard parenthood is (and thereby scare them off sex for ever, in my stereotyped picture of American high schools, although I suspect that's not fair).

I thought at first it was a weird affectation, but she turned out to be running a rather sensible experiment. She and Scott were considering having children, and she wanted to know what the disruption to her life would be like. So she got one of those babies that wails when you leave it alone, and wakes up several times in the night, and needs its nappy changing, and so on, to get an impression of whether motherhood was for her.

What she had failed to consider, which I think is sweetly typical of the Rationalists in a lot of ways, is that people would stop her in the street and say, 'Oh, cute, a baby!' and she'd have to awkwardly explain that it was not, in fact, a baby, but an experimental robot. Still, I hope the experiment gave her some non-zero level of insight into what parenthood involves; I did try to downplay her expectations on that front, given my experience that the challenges of newborn-baby-parenthood are only vaguely related to those of hyperactive-toddler-parenthood and, I assume, even less related to the stages after that.

Another thing that interested me was the almost complete absence of small talk – I'm a nervous talker, so I found myself gabbling to fill spaces in the conversation. It was Big Topics or nothing. And they actually pay attention to the arguments you're making; in my incoherent blather I was trying to justify the idea of writing this book (of which they're all sceptical, to a greater or lesser degree), and used several, mutually incompatible reasons for doing it. Katja in particular noticed and pulled me up on it.

When the time came to pay for our pizzas, we played a strange little game. We used someone's phone to come up with a quasi-random number between 0 and the price of the bill, and then counted down through the items to see whose meal it ended up on; that person then had to pay the whole bill. (Imagine there were two of us, just me and Scott, and the bill said 'Tom's Pizza, $10; Scott's Pizza, $10'. If the random-number generator came up with a figure between 0 and 10, I'd have to pay for it, because I was first on the bill; if it came up with a figure between 10.01 and 20, Scott would have to pay for it.) It ended up on Katja, so she paid for everyone's meal. I felt guilty and tried to pay anyway, but Buck stopped me: 'If you'd lost you'd have had to pay the whole thing. It's fair.'

It has only just occurred to me now, as I'm writing this nearly eight months later, how clever this system is. Splitting a bill according to who had what is time-consuming, boring and socially awkward. But splitting it clean down the middle incentivises people to order more expensive meals; they get all the benefit of the nicer meal but only pay a fraction of the extra cost, in a classic tragedy of the commons. But this system was extremely quick – almost as quick as dividing the bill on a calculator – and people who ordered more expensive things were more likely to pay the whole bill: if you ordered lobster thermidor for $80, the chances would be much higher that the number would end up on your bit. On average, you pay for exactly what you ordered. (Although it would take dozens of meals out for the averaging effect to cancel out the random noise.) Anyway, I thought it was clever, and very Rationalist.

I went to a London meet-up as well, which was pleasingly British; it was in a pub, and some people other than me were actually drinking alcohol. (In California, with Scott and the others, I had one beer; they were all on Diet Coke, and although none of them was remotely judgemental about it I still felt like some great lumbering hooligan, as though I were about to rip my top off and start throwing the chairs around like an England football fan in a provincial Portuguese town.) There were 16 people there, mostly but not exclusively men, crowded around a table for perhaps eight in a Wetherspoons near Holborn. A large majority of the time was spent in a rather AGM-style discussion, earnestly establishing rules for how conversations should go: should there be set topics? Should there be reading materials?

This seemed to go on for an awfully long time, with no immediate danger of stopping, so I ended up following one of the more normal people to the bar and asking him what he enjoyed about it all. He lived outside London and it cost him £30 to get there, he said, so he didn't do it often, but 'I like being able to come here and not be normal, before I have to go home and be normal again.' What did he mean by that? I asked. He meant he could talk about weird topics – AI, transhumanism, existential risk, biases, all the weird, nerdy stuff – and reliably be among people who wouldn't think he was weird for doing it. 'Plus,' he said, 'I can be a bit of a dick, and I like that I can say something really controversial, and instead of them being offended they all lean forward and say, "Let's unpack that."' He did proceed – fairly shortly afterwards, once the AGM business had died down – to say something quite controversial about transgender rights, if I recall correctly, and lo, they did lean forward and they did talk about it seriously.

LessWrongers also began various IRL projects. The Singularity

Institute, now MIRI, already existed when the LessWrong decline began, but the Center for Applied Rationality (CFAR) was founded by Anna Salamon, Julia Galef, Valentine Smith and Andrew Critch in 2012.

Other projects grew out of the LessWrong diaspora. There's a food company called MealSquares which produces savoury cake-like things that are supposed to be a healthy meal in a polythene wrapper; the idea behind them is to make healthy, 'optimal' eating easy. (I tried one at CFAR's office; it was like a heavy, dry-ish scone. It was 'nice in an overpoweringly dense sort of way', according to my notes from the time.) There's Beeminder, a 'reminders with a sting' goal-tracking project that is free as long you keep hitting your targets; you only start *paying* for it if you miss them. You might want to start running 10 kilometres a week: you link your FitBit to Beeminder and it will charge you when you fail. It's honestly clever.

They haven't all worked out: there was a medical-consultancy start-up called MetaMed, founded by Michael Vassar, where Scott worked for a while, which tried to use the LessWrong Rationalist techniques in medical assessments; it failed rather sadly after a couple of years. I read a couple of post-mortems[1] online,[2] and the reasons it failed sounded very much as Robin Hanson would have predicted: the creators thought they could rebuild healthcare almost from scratch, and didn't realise that there were huge amounts of unspoken local knowledge involved in both healthcare and business. I don't want to sound mocking, but there were a lot of lines in the post-mortems which could be summarised as 'Turns out marketing is important' and 'Turns out you actually need to do the job in front of you as well as think about the glorious future where AI solves everything.' Robin's comment about everyone being smart and articulate and convinced they can do everything better than it's been done before, without really checking *how* it's been done before, seemed very apt here.

There was a no-one-knows-exactly-what-but-it-appears-to-have-been-a-sort-of-Wikipedia-for-maths project called Arbital. Eliezer Yudkowsky was involved in that, along with some other Bay Area Rationalists. Again, the problem seems to have been a surfeit of big ideas and a shortage of actually knowing what they wanted to be doing *right now*. It had a '55-page document describing Arbital and how and why it was different and necessary', written by Yudkowsky; it was a 'better Wikipedia', but it was also a blog and a discussion board and had a Reddit-like karma-upvoting system, and also provided ratings for things so it could replace services like Yelp. The post-mortem, after explaining all of this, sighs: 'Now you can probably see how the meme of "Arbital will solve that too" was born . . . we just didn't have a good, short explanation of what Arbital was.' It shut down in 2017 without ever really getting off the ground, although its legacy is a genuinely good intuitive explanation of Bayes' theorem (it's all still online).

I don't want to be too harsh about any of this. Most start-ups fail, a fact of which the Rationalists are extremely aware; one of the biases discussed in the Sequences is the tendency to overestimate our own likelihood of success. Taking the 'inside view', most people think their business will succeed, but taking the 'outside view', about half of new businesses fail in their first five years; that figure is higher for Silicon Valley start-ups, with their whole 'fail-fast' ethos, at between 60 per cent and 90 per cent, depending on which study you read. So there's absolutely no shame in two of the four Rationalist-led start-ups I am aware of failing in their first few years, and even though I do think it's interesting that they did so in (what seemed to me) quite predictable ways, Yudkowsky would no doubt point out that, with hindsight bias, everything seems obvious once it's happened.

Not all Rationalists spend time in the IRL community, obviously. Some fairly central members have rarely if ever met their fellow Rationalists in person. Jim Miller, a professor of law and economics at Smith University in Massachusetts, and one of the original Overcoming Bias crowd – he was blogging on Robin Hanson's website back in 2007 – spoke to me over Skype and said he'd never actually met any other Rationalists IRL, apart from, coincidentally, at a conference in Sweden a few weeks previously. 'If I was a college student I would have seriously considered moving to the Bay Area,' he said, 'if I wasn't married or anything. But not at this stage of my life [he's in his early fifties, and married with children]. Also,' he added thoughtfully, 'living in a small part of a house in Berkeley just seems kind of horrifying to me, actually. When I was living in a dorm it would have been fine, but going back to that kind of life . . .'

But even just online, the Yudkowskian project has nonetheless had a profound impact on his life. For one thing, he steered his career towards it: he teaches about the economics of the far future, talks in his classes about cognitive bias, and has written a book on the potential rise of AGI, called *Singularity Rising*. 'Before [Less-Wrong], I was a traditional economist,' he said. 'Not assuming that rationality is perfect, but thinking it's a pretty good model and we don't need to go beyond that. The LessWrong stuff convinced me that economists should be looking at cognitive biases.'

His day-to-day decisions in his personal life have changed to some degree, as well. Most dramatically, he is taking steps to extend his life, for a very simple reason: if the singularity comes, then the difference between dying the year before it and the year after it is almost incalculable. 'If we do achieve the singularity and allow indefinite life extension, it could easily happen after I naturally die, so the expected value to me of living a few more years is huge.' If there's, say, a 0.1 per cent chance that those extra few years might get him to the glorious future, and the glorious future means a subjective life of a million years, then that 0.1 per cent chance translates to an expected value of a millennium of extra life.

He has signed up for cryonics, as many Rationalists have, and is an adviser to the board of Alcor, one of the largest cryonics firms. He has changed his diet – 'I've gone mildly paleo' – and talked his doctor into prescribing him a diabetes drug called metformin, because there's some evidence that it extends lifespan through an anti-cancer effect. Is the evidence solid, I asked him? 'The evidence is solid that it doesn't do harm and it might be good,' he said. 'So many people take it that a significant negative effect would be known.' (Sincere warning: please do not take this as any sort of medical advice.)

This online-only connection to the Rationalist community is probably the norm. The 2016 LessWrong diaspora survey found[3] that only 8 per cent of respondents 'regularly' attended meet-ups, and another 20 per cent had done so 'once or a few times'. Another question asking about 'physical interaction' with other LessWrong community members (for example, 'Do you live with other LessWrongers, or are you close friends and frequently go out with them?') found 7.6 per cent did 'all the time' and 12.5 per cent 'sometimes'. The large majority, at least according to this survey, have never met another Rationalist in person. The 2018 Slate Star Codex survey found similar results: only 10 per cent of respondents had ever been to a meet-up, and only 24 per cent of those still went to them regularly.[4]

But there is a hard core of Rationalists, perhaps a few hundred or a few thousand worldwide, who are more committed: who go to the meet-ups and live in the group homes, who (in many cases) financially support MIRI or other Rationalist groups, and who engage in polyamorous relationships. They're an unusual bunch of people, and they are centred around a few charismatic figureheads. This has led to accusations of them being a 'cult'. But are they?

Part Seven

Dark Sides

Chapter 34

Are they a cult?

'They're a sex cult,' says Andrew Sabisky.

Sabisky is an interesting man. He's a superforecaster, like Mike Story (whose massive dog crushed my feet), and exists in the same internet circles as the Rationalists. Unlike Mike, though, he seems to be a sort of enemy of the Rationalists, or at least a thorn in their side – he's certainly one of their most vocal critics. He's also kind of weird, in an endearing sort of way. Though there's no doubting the sincerity of his beliefs, he seems to have decided to become a Christian on the basis that tradition and ritual are important for humans, started running the social-media accounts for a central-London church, and *appears* to have got from there to actually believing in God. ('He memed himself into it,' Mike Story says, fondly. Apparently, there's a theological idea of 'act as if you have faith and faith will be given to you', and that's what Sabisky did.) During our two-hour conversation in a swanky central-London coffee house, we strayed away from the Rationalists and AGI and at one point ended up talking about the English Civil War, which Andrew is firmly against. 'They cancelled Christmas and they killed our king!' he declares, in a loud and declarative voice that may be ironic but may just be wearing the clothes of irony to distract from the fact that he 100 per cent means it.

The idea that the Rationalists are a cult – whether sex or otherwise – is not uncommon. It is, in fact, the subject of much writing by Yudkowsky and Scott Alexander, who appear to worry about it as a realistic problem. In fact, in the process of writing the Sequences Yudkowsky wrote so many things about avoiding cultishness that the word 'cult' started getting suggested as an autocomplete when you searched for LessWrong on Google. This was seen as suboptimal, so (in a delightfully Rationalist way) they asked everyone using the word 'cult' in a post to use instead a simple substitution code, shifting all the letters by 13 in the alphabet – so A→N, B→O, etc. – and turn it into 'phyg'.[1] (It doesn't appear to have worked: when I search 'lesswrong' in an incognito window so my search history doesn't come up, 'less wrong cult' is the second option. Still, it was a nice idea.)

It's worth addressing. They do share a lot of the surface features of a cult: a charismatic figurehead and other high-status inner-circle members; a key text that in-group members are supposed to have read, and which encodes the central tenets of their 'belief'; unorthodox sexual practices; a message of impending apocalypse, and a promise of eternal life; and a way to donate money to avoid that apocalypse and achieve paradise.

In case that needs spelling out, I mean Yudkowsky as the figurehead and others – Bostrom, Scott Alexander, Rob Bensinger, Luke Meulhauser, etc. – as the high-status inner-circle men; the Sequences, and to some extent Yudkowsky's Harry Potter fanfic *Harry Potter and the Methods of Rationality*, as the local bible; the Rationalist tendency towards polyamory as the sex stuff; the AI apocalypse, and the cosmic endowment, as the eternal life; and MIRI as the recipient of the indulgences.

'Yudkowsky and his wife and girlfriends,' says Sabisky. 'He used to be, like, an uber virgin. Then he got famous and started a sex cult, as you do. What else would you do with all that fame within a very narrow circle? If you look at his output, the main one is the Harry Potter fanfic. It's not aimed at making money, it's clearly just a thing that attracts people into your sphere. That's their whole point.'

'They'd have to be fucking blind to see they hadn't formed a cult', according to David Gerard of RationalWiki, which in the internet's sceptical–rational website wars is the Judean People's Front to LessWrong's People's Front of Judea. 'They tried not to become a cult, they asked themselves if they were being cultish, which was the right thing to do, but it happened anyway.' The Effective Altruism movement is part of it, he says, insofar as it wants you to give money to prevent AI apocalypse. 'Some charities are more effective than others, and you should donate to the more effective ones,' he wrote on Tumblr, 'and clearly the most cost-effective initiative possible for all of humanity is donating to fight the prospect of unfriendly artificial intelligence, and oh look, there just happens to be a charity for that precise purpose right here! WHAT ARE THE ODDS.'[2]

I asked Paul Crowley about the whole cult thing, and it was the one time he got even slightly angry. 'It drives me up the wall!' he said. 'We're exceptionally good at this.' By 'this' he means 'not becoming a cult'. 'But there's no way to say, "Oh no we're not", because you must be a cult if you're denying it.' He acknowledges that the Rationalist movement is 'weird': 'There's no way to deny that.' But they have a cause, and that cause is saving the world from unfriendly AI, and so they want to get people involved with it. 'Everyone does that, right?' he asked. 'You think, *you know what, climate change is a problem!* So I need to get some people on board with the idea that climate change is a problem. Or *I think the Democratic Party is going in the wrong direction, I'll get everyone I know on board with changing it.* As soon as you have a cause, you want to get people on board and say we should talk about this.'

According to Paul, the thing that distinguishes a cause from a cult is when it becomes taboo to criticise the cult. 'What's danger-ous is when you start to attack people's ability to think critically about it,' he said. 'A common trick, for example, is to say that questioning the precepts of this is morally wrong. If someone says, "I'm not sure that's true," and your reaction is, "You're a bad person for even asking," then that starts to get dangerous. On that score, I think we do unbelievably well! We're out there on the far end of the scale of how comfortable we are with people asking those questions.'

This is something that Yudkowsky has thought about himself. He thinks cultishness is 'a high-entropy state into which the system trends, an attractor in human psychology', by which he means that, left to its own devices, any group based on some noble cause will naturally slide into something like a cult. 'Every group of people with an unusual goal – good, bad, or silly – will trend toward the cult attractor unless they make a constant effort to resist it. You can keep your house cooler than the outdoors, but you have to run the air conditioner constantly, and as soon as you turn off the electricity – give up the fight against entropy – [it] will go back to "normal".[3]

It doesn't matter, writes Yudkowsky, if the cause is one of rationality and science and introspection: 'Labelling the Great Idea "rationality" won't protect you any more than putting up a sign over your house that says "Cold!". You still have to run the air conditioner – expend the required energy per unit time to reverse the natural slide into cultishness. Worshipping rationality won't make you sane any more than worshipping gravity enables you to fly.'

He has dedicated quite a lot of time and energy, in the form of several blog posts of some length, to avoiding the problems of cultishness. The question is: how well have they succeeded?

A lot of the LessWrongers do hero-worship Yudkowsky, to some degree. You can read it in the tone of their posts. That's hardly surprising, since he's a charismatic figure at the centre of everything they care about, and many of them are young men and teenage boys who have a tendency to do that sort of thing. One woman, who is more involved in the Effective Altruist community but hangs around with Rationalists as well, told me (anonymously): 'I do think the Rationalist community, at least in parts, has troubles with this heroic narrative, where some people are these sort of superheroes who can do anything. Some people see Eliezer as that, although I don't know if he plays up to it. And [another high-profile Rationalist] sees himself like that.'

And there's the fact that they want you to donate money to MIRI to stop the world from being destroyed. That is *quite* apocalypse-culty. Take this, for instance, on the money side of things: Yudkowsky was asked in 2010 what his advice was for people who want to help save the world, and he said: 'Find whatever you're best at; if that thing that you're best at is inventing new math of artificial intelligence, then come work for the Singularity Institute. If the thing that you're best at is investment banking, then work for Wall Street and transfer as much money as your mind and will permit to the Singularity Institute where [it] will be used by other people.'[4] It's not great, at first blush, is it? 'Get rich and give us every penny you can spare to prevent the apocalypse.' And people did donate (although a difference from a classic apocalypse cult is that donations are intended to save *everyone*, rather than to buy salvation for the donor specifically).

That said, there is a pretty good counterpoint against the idea that the Singularity Institute, as it was then, or MIRI as it is now, is a hoover for sucking up gullible people's donations. That is, Yudkowsky has stopped asking for them. He appeared on the philosopher Sam Harris' podcast in early 2018, and stated: 'thanks mostly to the cryptocurrency boom – go figure, a lot of early investors in cryptocurrency were among our donors – the Machine Intelligence Research Institute is no longer strapped for cash, so much as it is strapped for engineering talent'. He feels that MIRI needs more engineering staff in order to spend the money it has. 'We can obviously still *use* more money,' he said when I asked him about it, 'but our organisational attention has shifted to finding researchers and engineers.'

And, according to their finances, which are available on their website, the 10 'research' staff shared a total annual salary of $585,000 in 2017.[5] No doubt Yudkowsky and the executive director Nate Soares account for a decent percentage of that, but even if the other eight people are on a fairly-low-for-Berkeley $40,000 a year, it would put those two on about $130,000, which is obviously a decent wage, but pretty comparable to that of a good software engineer. It's not a perfect metric, but I would imagine that Joseph Smith wouldn't have been quite so abstemious in paying his own wages out of the Church of Jesus Christ and Latter-Day Saints' revenues.

But here's why I don't think they're a cult. Or, actually, let me put it another way. You could call them a cult, if you like. But it would involve defining the word 'cult' in terms that would remove most of the things about cults of which we are most wary.

I'm going to expand on this by referring to a post of Scott Alexander's, from before his LessWrong days.[6] It was titled 'the worst argument in the world'. The argument goes like this: 'X is in a category whose archetypal member gives us a certain emotional reaction. Therefore, we should apply that emotional reaction to X, even though it is not a central category member. Call it the Noncentral Fallacy,' says Scott. It sounds a bit obscure, but we all do it, all the time.

Scott's first example is Martin Luther King. Imagine someone wants to build a statue of MLK in a city somewhere. Someone objects to it: 'But Martin Luther King is a *criminal*!' What's your response to that? 'No he wasn't'? But he was: he broke the law by protesting against segregation. It was a shitty law, but he broke it. He was a criminal. And your opponent is saying that because criminals are bad, and MLK was a criminal, we should think MLK was bad.

'The archetypal criminal is a mugger or bank robber,' says Scott. 'He is driven only by greed, preys on the innocent, and weakens the fabric of society. Since we don't like these things, calling someone a "criminal" naturally lowers our opinion of them.' But MLK is a noncentral example of a criminal. He wasn't driven by greed or preying on the innocent. 'Therefore, even though he is a criminal, there is no reason to dislike King.' But this is a really hard thing to argue against, when you're in the brutal cut-and-thrust hand-to-hand combat of an argument on the internet. You would not instinctively respond, 'Yes it is true that Martin Luther King was a criminal, but he did not share the features of criminals that make them bad, so your suggestion that he is bad is based on faulty logic.' You would, Scott suggests, be much more likely to say: 'No he wasn't! Take that back!' And then the exchange becomes about whether he was a criminal or not, and 'since he was, you have now lost the argument'.

This sort of thing happens all the time. 'Abortion is murder', 'Taxation is theft'; the argument is, 'Abortion shares some features with murder, so you should be as outraged by it as you are by murder', or 'Taxation shares some features with theft, so you should be etc.' And of course, if you try to argue against it by saying, 'Well, it's the good kind of theft', your unscrupulous enemy will say (to quote Scott): 'Apparently my worthy opponent thinks that theft can be good; we here on this side would like to bravely take a stance against theft.'

This is what I think is going on in the 'cult' argument. I spoke to some Rationalists who were, in fact, much more sanguine about describing LessWrong and Rationalism as a cult, or a religion. For instance, Ben Harrison, the young British Rationalist who founded a group house in Manchester, was happy to call it exactly that. 'I was accused of being a cult leader last week,' he said, offhandedly. He's pretty laid-back about the whole thing. 'There are elements of religion to it. The structure it provides, the figures [such as Yudkowsky]. We've got holy texts, people who dedicate their lives to it. It has most of the trappings of religion; I'd call it a pseudo-religion. I can see what people are pointing at there.'

But where the 'worst argument in the world' comes in is that while it has a lot of the *features* of a cult, it is not a central member of the category 'cult'. 'The word "cult" is really just a name for a strong community that's disapproved of,' Robin Hanson – the economics professor who first hosted the Sequences on his blog, Overcoming Bias – told me. 'When I was a young teenager, 12 or 14, I was in a Christian cult.' But the cult never ordered him to dissociate himself from his friends or family, and when he left after a year or so, he was free to do so. 'There are coercive cults, but the cult I was in wasn't one, and [the Rationalist community] certainly isn't either. It's rare, really, for cults to have these dissociation rules. Usually they just have strong eagerness to get close to one another, and spend time with one another in group homes and so on.'

That's not to say there's nothing to worry about. Anna Salamon, the founder of the Center for Applied Rationality, told me that she thinks the 'cult' question is probably the wrong one. 'I don't worry about the cult thing very much,' she said. 'I agree it's important that we [don't coerce people or crack down on dissent].' She doesn't think they *do* do that, for the record. What she worries about is that people will miss the thing they're trying to do – teach ways of thinking, ways of examining your own thinking – and just pick up on the 'AI apocalypse' conclusions. 'I worry about making sure that our classes aren't too convincing of the wrong thing. We try to share the processes for thinking, and how to verify those processes, rather than duping people into believing specific conclusions because someone at the front of the room said them.'

The mental techniques – the use of maths and Bayesianism to support conclusions about everyday things, for instance – are weird and 'a little bit scary', and they try to get people to take them seriously. But if you just put, say, Bostrom's numbers about the cosmic endowment in front of someone, and don't teach them the stuff that contextualises them – that lets you check for yourself whether it's worth being worried about it – then, she thinks, it could go a bit wrong. 'It freaks me out a bit,' she said. 'I'm a little afraid that someone *will* just trust the numbers. For example, AI risk is a very big topic, and it's easy for people to feel hijacked by it, and be like, "Aha, I'll ignore all the things I normally care about and just care about this thing." I think that's mostly a bad idea.'

It's not that she doesn't want people to worry about AI risk, but rather that she wants people to be able to examine the question – the maths, and reasons why you shouldn't just blindly follow the maths – for themselves. She wants people to work that out for themselves, not to take it on faith from some authoritative source. This strikes me as doing the work that Yudkowsky mentioned, the 'air-conditioner' work: actively trying to stop your cause from turning into a cult. From my point of view, it seems to be working.

But Sabisky's suggestion wasn't simply that they are a cult – it was that they are a sex cult, that the entire movement exists to (or at least has conveniently turned out to) enable powerful Rationalist men to have sex with lots of impressionable young women. This revolves, at least in part, around the fact that a large number of the full-time Rationalists – the ones who live in group houses and/or work for CFAR or MIRI, and the ones centred around the FHI in Oxford – are polyamorous. This is true of both men and women. Eliezer Yudkowsky himself is polyamorous: he has a wife and two other partners, all of whom are themselves polyamorous and engaged in other relationships.

This is where the 'sex-cult' accusations come in. You can see how that might be something worth worrying about. A man builds a community in which he is a figurehead; he gains power and prestige which enables him to attract the sexual attention of several women; the community conveniently develops an unwritten rule which says that he is allowed to have sexual relationships with several women. It 'pattern-matches', to borrow the terminology that the Rationalists use, extremely well to things like David Koresh, the Texan cult leader and 'final prophet' who died under police gunfire in a siege in 1993. This is basically how Andrew Sabisky sees it. 'It seems to revolve around the highest-status men, and they get to pick and choose.'

But there are several ways in which this story is at best incomplete, and, at worst, downright wrong. For one thing, Paul Crowley is a Rationalist, and he's also poly, but he was poly for many years before he became a Rationalist; he's never been anything *but* poly. It's absolutely the case that some people did turn poly after joining the Rationalists – Scott Alexander, for instance – but to some degree, at least, the Rationalist community *attracted* poly people, rather than simply instilling a code of polyamory from on high.

Also, it should be pointed out that polyamory is not the norm even among Rationalists. It's more common in the Rationalist community than elsewhere, I think, but perhaps not as much more as you'd expect: the Slate Star Codex 2018 survey[7] found that slightly less than 10 per cent of its 8,000 respondents said their preferred relationship style was poly; a 2014 LessWrong survey[8] put the figure at about 15 per cent.

I don't have reliable numbers for the population at large, but I read an estimate (from a blog post in *Psychology Today*, quoting an 'independent researcher'[9] – apply salt liberally) that there were 'around 1.2 to 2.4 million' American couples 'actually trying sexual non-monogamy'; about 10 million if you include all couples who 'allow satellite lovers'. There are about 120 million adults of each sex in the US. If we assume that three-quarters of them are in committed relationships, then we're looking at something between 2 per cent and 10 per cent of all couples who are poly, depending on your definition. A study found that 21 per cent of people had at least *tried* it, at some point.[10] Another article in the legal magazine *Advocate* claimed, with no references, that 'most researchers estimate that a full 4–5 percent of Americans participate in some form of ethical non-monogamy'.[11] I think it's probably OK to guess that 5 per cent of Americans prefer polyamorous relationships. That makes it at least twice as common among Rationalists, but we're not dealing in orders of magnitude or anything.

Also, the average Rationalist is significantly younger than the average American. I have no data on this whatsoever, but since articles about polyamory tend to refer to it as a 'new trend', I assume it's more common among young people.

I expect it's even *more* common among people who tend to look at social norms and say: 'Why do we do this and do we need to carry on doing it?', which is of course exactly the sort of person who would join the Rationalists. 'I think there's a phenomenon where a social circle that is OK with unusual choices will have a lot of poly people in it,' points out Paul Crowley. 'Lots of folk I know on the goth scene are poly, for example. Same for kink, sexuality, transness.' It may well be that Rationalists are no more likely to be poly than any other group of young, nonconformist oddbods. The superforecaster Mike Story agrees: 'The polyamory thing isn't like, "Let's be in the Rationalists and have lots of sex." It just so happens that if you think *this* way, you probably also think *that* way.'

All that being said, the surveys are addressed to the wider internet population. There isn't some ringfence around 'the Rationalist community' which says this person is in and this person is out; everyone who read LessWrong in 2014, and everyone who read Slate Star Codex in 2018, was asked to take the survey. I think I took the 2017 SSC one, and although I'm a sympathiser, I'm not really a member of the community.

Ben Harrison, the founder of the group house in Manchester, pointed out to me when I spoke to him that there are circles and circles. 'The biggest is the people who read Slate Star Codex casually,' he says, a circle which includes a large number of people I know who absolutely would not call themselves 'Rationalists'. 'Then there are the wider people who read all the blogs but don't feel the need to be part of the scene socially. Then there are all the people on the Discords and the Slack channels [group chats].

Then there are the people who go to the meet-ups, then the people who live in the group houses, and then the people who exist, in an economic sense, entirely in the community.'

And it does seem that the further you get into the inner circles, the more likely it is that you'll be poly. That makes sense in the light of the idea above: the more devoted you are, the more nonconformist you're likely to be, compared to vanilla dilettantes like me. But I think a majority of the people who actually work at CFAR and MIRI, and (although no one will actually confirm this for me) large numbers of the Future of Humanity types are poly. Buck Shlegeris of MIRI, the young Rationalist I had pizza with, estimated that 70 per cent of Rationalists are poly. It depends where you draw your line around the term 'Rationalist'.

The poly aspect has undoubtedly led to scenarios which would raise a lot of people's eyebrows. 'Have you heard the story about the LessWrong baby?' Sabisky asks me, somewhat salaciously. I had, vaguely. What happened was that a young woman in the Bay Area was in a polyamorous relationship, or interlocking set of relationships – the standard mono terminology gets a bit confused. She had a husband, but she was also in a relationship with another man, a high-status, extremely wealthy member of the Rationalist community who himself was married.

'So anyway she's shagging this guy,' says Sabisky. 'And they have an agreement – because, *apparently*, contraception is hard to use – that if she gets pregnant she'll have an abortion. Well, she does get pregnant by him, and can't follow through, understandably.' The upshot is that she had the baby, who by all accounts is now doing very well; she broke up with her husband, although that sounds like it was less to do with the baby and more to do with issues of his own.

Where it all blew up a bit was that someone started a crowdfunding thing among the Rationalists to help support her. She had, admirably to my mind, refused to try to get money from the father, because of this deal that she'd made. Scott Alexander plugged the crowdfunder gently in one of his blog posts. 'And people slated her! *Slated* her!' says Sabisky. 'For not having the abortion. "You made a contract! You made a deal! Why should we pay for you?"'

I have checked, and there is indeed a lengthy thread in the comments pretty much along these lines, occasionally interrupted by the woman herself saying 'Please try to remember that I can read this and it is hurtful.' It really wasn't very pleasant. It made me understand, on a visceral level, exactly why some people dislike the Rationalist community: they're unsettlingly willing to discuss things that many people find sacred, in a way that those people find profane. Most of the time it doesn't bother me: I'm quite happy talking about the dollar value of a human life, say. But I found the open discussion of whether a woman ought to have aborted her now-18-month-old daughter because of a deal she'd made honestly hard to read. I'm a father of two, and perhaps it affected me personally in a way that other things do not.

But is it a David Koresh-style sex cult? Well. I think the 'Less-Wrong baby' story is unpleasant: a powerful and significantly older man driving his girlfriend to an abortion clinic, wanting no more to do with the baby once it was born, and not helping financially, because of a deal they had made before she got pregnant. But if this is the most egregious example of sexual abuse of power in the Rationalist community, then it seems, not trivial, but certainly no worse than things that happen all the time in many monogamous relationships.

The question, of course, is whether there are worse things. There's this godawful Reddit page called /r/sneerclub which is dedicated to mocking the Rationalists, and they latched with enormous glee onto an OKCupid profile that Eliezer Yudkowsky put up in 2012. If you've ever read anything by Yudkowsky it's immediately familiar: bombastic and self-promoting in a way that is meant to be ironic but isn't really. It's also very open about his fetishes, which attracted some mockery. And his 'you should message me if' bit says that 'my poly dance card is mostly full at the moment', but that you 'shouldn't worry about disqualifying yourself or thinking that I'm not accessible to you', and should instead 'test experimentally what happens when you try asking directly for what you want – that's Empiricism'. 'I'm also cool with trophy collection,' he adds, 'if you only want to sleep with me once so you can tell your grandchildren.'

It gives me an icky sensation, but again, it's just a bit weird, not wrong, at least as far as my moral framework goes. If women want to message him, great; I hope it works out for all parties. But for some people the 'ick' factor is stronger, I think. I've heard it described as Yudkowsky 'trawling for sex', but I can't see how you get there. He's just openly stating that he wants sex and that if anyone wants it with him they should ask.

I should also acknowledge that in June 2018, some extremely dark allegations were made in a suicide note by a former Effective Altruist and Rationalist who said she was abused. It is an awful read. But as far as I can tell, the allegations had already been properly investigated by people appointed to safeguarding roles in the Effective Altruism community. Some of the allegations were found to have a basis in fact, and some people were barred from, or had already been barred from, parts of the community; other allegations were found to have no such basis. You'd expect there to be some awful people in a community as large as the Rationalists, though, and it seems to me – having spoken to several people close to the issue – that the situation, though messy and complicated, was well handled by the people whose responsibility it was to do so.

Here's my position. I don't think the Rationalist community is a sex cult. But people on the *outside*, those of us like me in our hetero-monogamous,married-and-settled-down,two-kids-and-a-people-carrier world, find their whole thing deeply weird, and for us it is very hard to separate 'weird' from 'immoral'. For instance, for a lot of people it would be difficult not to be jealous. I asked Paul about this when I saw him and he said it just . . . never occurred to him. He is aware there are good evolutionary-psychology reasons why he should be jealous, but he's just not.

Dr Diana Fleischman, an evolutionary psychologist herself at the University of Portsmouth and a prominent member of the Effective Altruism movement, thinks that this can come across as really weird to other people. 'To the general public, it can seem like realigning your evolved motivation to such an extent that it makes you untrustworthy. If you can rewire your jealousy then you're really dangerous, because you're capable of anything. I think that's part of the stigma.' Fleischman is polyamorous herself, and completely unembarrassed discussing it, which makes one of us. She talked a bit about the 'harem' accusation. She pointed out that the arrangement tends to be genuinely *polyamorous*, not *polygynous* – it's not that the men all have lots of girlfriends but the women are expected to remain faithful. All of the high-profile men do have multiple girlfriends, but those girlfriends usually have several boyfriends themselves.

This actually solves a problem, in her view. The Effective Altruist and Rationalist communities are heavily gender-biased: they're mainly men. 'Polyamory fixes the numbers problem,' she said. 'I made this joke once: Effective Altruism is like 75 per cent male, but the 25 per cent of women all have three boyfriends.' I asked her whether it was a sex cult, and she laughed. 'I was joking with someone the other day. "The Rationalist community isn't just a sex cult," we were saying. "They do other great things too!"'

Chapter 35

You can't psychoanalyse your way to the truth

There's another point, which is that (on one, quite important, level) it doesn't matter whether the Rationalist community is a cult, or a religion, or not.

Take another example of a thing which some people say has the hallmarks of a religion: environmentalism. People who worry about the environment warn of an encroaching apocalypse, in the form of climate change. There is a prelapsarian past from which we have fallen away, through our own sin; it has charismatic prophets (such as George Monbiot and Al Gore) who warn of the impending doom; it has rituals we can perform in order to prevent that doom (such as recycling, or driving a Prius).

I should say, by the way, that I've lifted that example from Scott Alexander, who was responding to a specific person claiming that environmentalism is indeed a religion.[1] He pointed out that, also, people have suggested that liberalism, conservatism, libertarianism, the social-justice movement, communism, capitalism, objectivism, Apple, and the operating system UNIX are also all religions. You can, depending on where you draw the line around the category 'religion', include pretty much anything in it.

For one thing, this is another example of the 'worst argument in the world'. If I want to tar something with the brush of 'religion' (assuming that I am someone who thinks that religion is a bad thing), then I can point to these characteristics, claim that they are part of the definition of 'religion', and then say, 'Therefore, environmentalism or whatever is a religion and you shouldn't like it.' But, even more importantly, *it doesn't tell you whether environmentalism is right or not.*

Let's say that environmentalism really does share all the psychological hallmarks of real old-time religion. Say that, actually, the local Oxford low-carbon group or the cycling-promotion work that my extremely green parents are involved in are 100 per cent functionally identical, psychologically speaking, to churchgoing or evangelising. Say that paying for carbon offsets does precisely the same thing, in some religion-shaped part of your brain, as buying indulgences did for pre-Reformation Catholics. Say that the 2015 United Nations Climate Change Conference in Paris was indistinguishable, in its mental role, from the Second Vatican Council. Does that mean that carbon emissions are not in fact heating the planet? No! It tells us *almost nothing at all.*

Whether or not the world is warming up is a fact that is based on how much heat energy there is in the atmosphere and the oceans. The reasons behind whether or not someone *believes* that the world is warming up is a fact about people's brains. 'You can't find out whether the world is warming by asking about the psychology of Greens,' Paul Crowley grumbled when I put the whole 'cult/religion' thing to him. 'You have to look at satellite data, and CO_2 concentration, and this kind of thing. It's ultimately a massive distraction. Psychology just doesn't work if you want to find out about the world.'

Eliezer Yudkowsky has addressed this before, in an interview with the science writer John Horgan, who had previously called singularitarianism a 'religious rather than scientific vision'.[2] 'You're trying to forecast empirical facts by psychoanalysing people,' he told Horgan. 'This never works. Suppose we get to the point where there's an AI smart enough to do the same kind of work that humans do in making the AI smarter; it can tweak itself, it can do computer science, it can invent new algorithms. It can self-improve. What happens after that – does it become even smarter, see even more improvements, and rapidly gain capability up to some very high limit? Or does nothing much exciting happen?'[3]

Answering that question requires knowing actual facts about computer science, says Yudkowsky. You need to *go and look.* You cannot answer it by looking at the people who believe it and seeing whether you like them, or whether they seem weird.

So the question 'Is the Rationalist community a cult?' is not without value, insofar as would-be Rationalists might be in danger of getting fleeced of cash or manipulated into sex. But it doesn't address the underlying question, which is: 'Might humanity be destroyed by a badly aligned AI?'

That question is about what intelligence is, and how likely we are to be able to build it, and, if we do, what characteristics it will have (or we will give it). I believe there are excellent reasons to think that it might never happen (as well as some excellent reasons to think it might happen sooner than we'd expect). Not one of those reasons, though, is 'because those guys on LessWrong look a bit culty to me'.

Chapter 36

Feminism

There's another thing I am just going to have to address, which is: are the Rationalists a bunch of scientific racists, Trump voters, alt-righters and misogynistic Men's Rights activists?

As I've said before, I *like* the Rationalists. I think they're a well-meaning and interesting lot. So it won't surprise you to know that the short version of my answer to the question is going to be 'No'. The slightly longer answer is going to be 'No, but I can see why you might think that.' The really long answer is below.

The Rationalists are nerds. They are, usually, people who are deeply interested in how things work. They are disproportionately male, and disproportionately on the autistic spectrum or near to it, according to the LessWrong and Slate Star Codex surveys – accordingly, as autism usually involves social deficits, many of them lack social skills to some degree or another.

Lots of them are, you will be unsurprised to hear, not very good at talking to women. This blew up in spectacular style at the end of 2014. Scott Aaronson is, I think it's fair to say, a member of the Rationalist community. He's a prominent theoretical computer scientist at the University of Texas at Austin, and writes a very interesting, maths-heavy blog called Shtetl-Optimised.

People in the comments under his blog were discussing feminism and sexual harassment. And Aaronson, in a comment in which he described himself as a fan of Andrea Dworkin, described having been terrified of speaking to women as a teenager and young man. This fear was, he said, partly that of being thought of as a sexual abuser or creep if any woman ever became aware that he sexually desired them, a fear that he picked up from sexual-harassment-prevention workshops at his university and from reading feminist literature. This fear became so overwhelming, he said in the comment that came to be known as Comment #171, that he had 'constant suicidal thoughts' and at one point 'actually begged a psychiatrist to prescribe drugs that would chemically castrate me (I had researched which ones), because a life of mathematical asceticism was the only future that I could imagine for myself'.[1] So when he read feminist articles talking about the 'male privilege' of nerds like him, he didn't recognise the description, and so felt himself able to declare himself '"only" 97 per cent on board' with the programme of feminism.

It struck me as a thoughtful and rather sweet remark, in the midst of a long and courteous discussion with a female commenter. But it got picked up, weirdly, by some feminist bloggers, including one who described it as 'a yalp of entitlement combined with an aggressive unwillingness to accept that women are human beings just like men' and that Aaronson was complaining that 'having to explain my suffering to women when they should already be there, mopping my brow and offering me beers and blow jobs, is so tiresome'.[2]

Scott Alexander (*not* Scott Aaronson) then wrote a furious 10,000-word defence of his friend.[3] I can't begin to do the argument justice, and in all honesty I'm scared of doing so, because I don't want this book to be described as 'an anti-feminist screed'. But his point, I think, can be boiled down to the idea that 'male privilege', while a real thing, is not the only form of privilege, and that a lot of online discourse comes down to attempts to define group X as having *more* or *less* 'privilege' than group Y, as though privilege is a one-dimensional thing. (Hence the constant battle over whether trans women have male privilege.) So if nerds claim to be suffering, then, to the people who think of privilege on this one-dimensional axis, they must be saying that they are *less* privileged than women.

There's also a point he makes, which I think is a fair one, which is that nerdy people *really are* bad at talking to members of the opposite sex, and that this is a way in which they are truly disadvantaged: that nerdy people really do struggle with finding mates, that nerdy men are much more common than nerdy women (and men in general find it harder to attract sexual partners),[4] and that love, sex, intimacy and affection are important things in human lives, without which we are usually less happy. This is a real problem that real people face. Furthermore, nerds are likely to be bullied and abused for *being* nerdy (I often think a lot of the sneering comments online about 'fedoras' and 'neckbeards', common terms of abuse on the internet for socially unskilled young men, are little more than an extension of playground bullying of autistic and/or nerdy children), and you have a genuinely bad situation which causes real pain.

And you can say all that without saying that nerds are *less privileged* than women, or that their suffering matters more, and while acknowledging that nerdy men find certain careers easier to get into and progress in than nerdy women do. Instead you can just, in Scott Alexander's words, say: 'You feel pain? I have felt pain before too. I'm sorry about your pain ... I will try to help you with your pain, just as I hope that you will help me with mine.' Anyway, inevitably enough, Scott Alexander's blog post defending Scott Aaronson blew up and everyone accused Scott Alexander *as well as* Scott Aaronson of being a sexist entitled nerd.

There's a related problem, which is the 'women in science, technology, engineering and maths [STEM]' thing. It's true that women are comparatively rare in some STEM careers, especially engineering and computer science; a study looking at 1998 data found that only 26 per cent of American tech workers were female.[5] (Some reports suggest that the figure has actually gone down since then, to perhaps 20–23 per cent.[6]) Where the Rationalists (and specifically Scott Alexander, again) have got themselves in trouble is by suggesting that this might have other causes as well as simple discrimination; that it's not purely that Silicon Valley techbros are keeping women out.

Scott suggests, firstly, that if sexism was the key driver in keeping women out of Silicon Valley, you'd expect to see lots of girls doing computer science at high school but then not making it into tech careers. But that's not what happens: about 20 per cent of children taking high-school computer-science classes in the US are girls.[7] (About 8 per cent of A-level computer-science students in the UK are female.[8]) A 1989 study found a similar pattern at middle-school level.[9]

And it's not that sexist stereotypes are making women believe themselves to be worse at computer science than men and stopping them from taking computer-science classes, he argues, because surveys show that women *don't* believe themselves to be worse at computer science than men.[10]

Instead, he suggests, a major reason why women don't end up in Silicon Valley isn't that Silicon Valley doesn't want them – it's that they don't want Silicon Valley. Women are less interested, according to meta-analyses,[11] in 'thing-oriented' careers, and more interested in 'people-oriented' ones. 'I would flesh out "things" to include both physical objects like machines as well as complex abstract systems,'[12] says Alexander. 'I'd also add in another finding from those same studies that men are more risk-taking and like danger. And I would flesh out "people" to include communities, talking, helping, children, and animals.'

And, he says, this predicts things pretty well. For instance, medicine was once the absolute bastion of male privilege; about 50 per cent of US medical students are now female. (He doesn't give a source for this, but the UK figure is that 55 per cent of medical students were female in 2015,[13] and 56 per cent of students in medicine and dentistry in 2016–17.[14]) But, interestingly, when you dig deeper, you see that 'thing-oriented' branches of medicine – the ones where you don't talk to patients much but chop them up, or anaesthetise them, or blast them with radiation; the ones where people are treated more as objects, or systems, rather than people – are significantly male-oriented: surgery, 59 per cent male (in the US); anaesthesiology, 63 per cent; radiology, 72 per cent. And more 'people-oriented' branches – psychiatry, 57 per cent; paediatrics, 75 per cent; family medicine, 58 per cent; obstetrics/gynaecology, 85 per cent – are dominated by women.[15] Computer science is very thing-oriented, so you'd expect to find more men pursuing careers in it. The story appears to be similar in the UK: according to the General Medical Council, more than 50 per cent of British obstetrics and gynaecology specialists are female (and 66 per cent of those under 40 and 78 per cent of trainees), compared to just 12 per cent of surgeons.[16]

Of course, this doesn't rule out the possibility that the differing interests of men and women are entirely caused by socialisation in the child's early years. There's a lot of research on that, and endless back-and-forth arguments, and it's an enormous can of worms that I absolutely don't want to open here, beyond agreeing with Alexander that it's 'probably our old friend gene-culture interaction, where certain small innate differences become ossified into social roles that then magnify the differences immensely'.[17] But even if there is no genetic input whatsoever (which is unlikely), early-years socialisation is, says Alexander, not something you can entirely blame Silicon Valley or nerds for.

This is an ongoing argument; some scientists think the people-vs-things dichotomy explains a lot of the gender split, others think it doesn't. But the trouble with coming out in favour of one side or the other is that, as Yudkowsky previously observed, debate is war, and arguments are soldiers. 'Sexism is keeping women out of tech' is an argument on 'our' side, the side that wants women to be able to do any job they want. So people who say, 'The fact that women are under-represented in tech could be largely due to systematic differences in male and female interests' are attacking one of *our* 'soldiers', and will often therefore be treated as though they're saying, 'Women are not meant to be computer programmers.'

You can see this war-soldiers stuff going on, I think, in the case of the 'Google memo' written by James Damore. Damore's memo suggested that the under-representation of women at Google was in part because of interests, rather than discrimination. He was fired from his job and called a Nazi[18] and a fascist;[19] my own feeling is that this was a huge over-reaction, especially since he explicitly said that 'Many of these differences are small and there's significant overlap between men and women, so you can't say anything about an individual given these population level distributions,' and 'I'm not saying that diversity is bad, that Google or society is 100 per cent fair, that we shouldn't try to correct for existing biases, or that minorities have the same experience of those in the majority.'[20]

Cordelia Fine, a professor of psychology at the University of Melbourne and author of three feminist-inspired books about neuroscience, told the *Guardian* that, while Damore had a tendency to over-emphasise the evidence suggesting innate tendencies, his memo was 'more accurate and nuanced than what you sometimes find in the popular literature' and full of ideas that are 'very familiar to me as part of my day-to-day research, and are not seen as especially controversial'. She felt 'pretty sorry for him', she added, and found it 'extraordinary' that he had been fired and shamed.[21]

But when you think in terms of debate and war, and arguments and soldiers, it makes perfect sense. The *specifics* of Damore's memo weren't that relevant. The point was that he was arguing something that looked as if it would give succour to the Donald Trump/Brexit/alt-right 'side'. Most of us would have realised that, I think, and been wary of sending that 'memo' to everybody at Google. But for a certain kind of mind, which many Rationalists have and as, apparently, James Damore has, it was not obvious.

It won't surprise you to learn that Damore is on the autistic spectrum. It also won't surprise you to learn that, after he was fired, the alt-right swarmed around him, inviting him on their shows. Milo Yiannopoulos interviewed him. He was now providing soldiers for *their* army.

Chapter 37

The Neoreactionaries

Another accusation is that the Rationalists are linked to the alt-right. And there is a very good and specific reason to think that they might have those links. That is that they do.

The online group known as the 'Neoreactionaries', which is a sort of strange medievalist subset of the alt-right, grew out of the Rationalist movement to some extent. They even left LessWrong and founded their own website, named (spot the reference) 'More Right'. Mencius Moldbug, the founder of Neoreaction, wrote a few blogs on Robin Hanson's Overcoming Bias before LessWrong split from it. Michael Anissimov, another prominent Neoreactionary, was until 2013 MIRI's media director. The pseudonymous 'Konkvistador' is a regular Slate Star Codex commenter.

They're a small and weird subculture. Mike Story said to me that it was best to think of it like this: 'LessWrong is kind of a social club for people with mild, high-functioning autism, or nerdiness.' But within that wider group there's a division, which Mike calls 'the people who have [those social deficits] but are kind of arseholes, and the people who kind of have that but are good.' The latter make up the Effective Altruism crowd; the former, the Neoreactionaries. I don't want to make it sound as if there's an equivalence here. About 20 per cent of Rationalists identify as Effective Altruists; 0.92 per cent as Neoreactionaries.[1] But the latter, while relatively rare, do exist, and have influenced the development of the Rationalist community.

According to Neoreactionaries, the world has been moving steadily to the left for several hundred years, and that it has correspondingly become less safe, less happy and less clever, and that it is impossible to speak your mind freely unless you toe certain leftist lines. They also believe that some ethnic minorities' poorer life outcomes – in education, income, crime, mental health, etc. – are due to biological and/or cultural factors *within those minorities*; that women are happier in more traditional, 'sexist' societies; that immigration from some developing-world countries actively worsens America by bringing in people with different, and worse, values. And, most notably, they think that democracy should be replaced by an omnipotent and unelected king.

It's not that these Neoreactionaries are completely separate from the rest of the Rationalists. If you spend any time in the comments on Slate Star Codex, you'll find them quite often; same on the SSC subreddit.

Here's part of what I think is going on. First, comment sections are the literal worst. There's a thing on the internet called the '1 per cent rule', which is roughly that there's a hyperactive 1 per cent in any internet community which creates the vast majority of whatever the content is (comments, Wikipedia edits, YouTube videos) while most of the rest just lurk, reading and watching. And, indeed, only 1.2 per cent of LessWrong readers surveyed said they'd commented more than once a week for the last year, while 84 per cent said they hadn't commented at all;[2] the Slate Star Codex survey finds similar. And of course, that hyperactive 1 per cent will disproportionately include the most ideologically driven. So judging the Rationalists by the ugliest comments you can find in the SSC subreddit is rather unfair.

Second, the Rationalists have a particular problem which is that their *whole thing* is taking opposing arguments seriously – what Alexander calls the 'principle of charity'. It is part of SSC's ethos that, 'if you don't understand how someone could possibly believe something as stupid as they do',[3] then you should be prepared to find that that's *your* failing, rather than theirs. And what that means is that if you're the sort of person who wants to go and talk about 'race science', for example, you'll find that going to a Rationalist website and doing so means that you aren't immediately blocked. Instead, you find people will talk to you seriously and engage with you.

And this is a noble and brilliant thing, in many ways! If you want to shout about how terrible The Enemy is, so that you get cheered on by Your Side, then you've got the whole internet in which to do that. But the Rationalist community is where you can speak, calmly and collaboratively, with people with whom you profoundly disagree, and try to change minds (and admit to the possibility that you will have your own mind changed).

Unfortunately, this means that you have to allow people in who say things you find completely appalling. But civil society in Britain and America, and I think much of the world, is getting measurably more polarised; its conservatives and its liberals just live in different worlds, talking to themselves about how terrible the other lot are. Places where they can talk *to* each other, like the Rationalist community, should be protected and encouraged, not reviled. When we're dealing with AI safety, which really could affect everybody on Earth and which – if the Rationalists are right – is an imminent and hard-to-avoid threat, it seems particularly good to have a place where ideas can be exchanged without people feeling they have to censor themselves.

There is another piece that Scott Alexander wrote. (I'm sorry to use 'Slate Star Codex' as a near-synonym for 'the Rationalists', but it is by far the most high-profile part of the movement, and the most overtly political.) It was called 'You Are Still Crying Wolf', written in the wake of Donald Trump's election. The piece argued that accusing Donald Trump of being 'openly racist' and 'openly courting the white-supremacist vote' was disingenuous, and was still 'crying wolf' in the way that accusing more mainstream Republicans like John McCain and Mitt Romney of being Nazis had been in previous elections. Alexander's point was that Trump had, at the very least, gone out of his way to *sound* non-racist, pro-LGBT, etc. He'd gone to black churches, claimed that it was his 'highest and greatest hope that the Republican Party can be the home in the future and forever more for African-Americans and the African-American vote', apparently spontaneously grabbed a rainbow flag out of the audience at a rally and started waving it, and waxed lyrical about his 'love for the people of Mexico'. Trump may or may not have racist attitudes, but calling him '*openly* racist' or '*openly* white-supremacist', said Alexander, was silly. If nothing else, he obviously wasn't being *open* about it.

A lot of people focused on it as Trump apologia (there being, in the left-leaning internet circles I exist in, nothing worse than being a Trump apologist, except perhaps being a Brexit voter). I think this is *spectacularly* unfair. A few quotes from Alexander on the topic: 'Please don't interpret anything in this article to mean that Trump is not super-terrible'; 'Trump is just randomly and bizarrely terrible'; Trump is an 'incompetent thin-skinned ignorant boorish fraudulent omnihypocritical demagogue with no idea how to run a country, whose philosophy of governance basically boils down to "I'm going to win and not lose, details to be filled in later."'

I don't know if he's right. But it seems a reasonable argument to make, and it *certainly* wasn't pro-Trump. Before the election, Alexander had urged his readers to vote for anyone *but* Trump (and specifically Clinton in swing states), because Trump was the candidate most likely to blow the whole world up in some stupid way.[4]

And it's not just that Scott is the smiling liberal face of a far-right movement. Yudkowsky, Bensinger and other prominent Rationalists expressed concerns about Trump. The Slate Star Codex 2018 survey[5] found that 29 per cent of respondents were registered Democrats, and just 9 per cent registered Republicans. (36 per cent of respondents were unregistered, and 21.6 per cent were non-Americans.) On a 'political-spectrum' question, asking people to rate their political position from 1 (far left) to 10 (far right), the largest single response was 3, at 26 per cent, followed by 4 at 21 per cent.

I think this needs saying, because the Rationalists frequently criticise what they call the 'social-justice movement' and which I tend to think of as the Twitter/Tumblr left. And if the Rationalists are alt-righters, misogynists and racists, then it's easy for the liberal left to disregard their criticisms. But if they're a group of largely liberal-left people, who have specific criticisms of aspects of mainstream liberal-left thinking, then it's harder to ignore them.

There's another point: a key argument that many Rationalists are making is that there is a real and believable risk that we might all be destroyed in the foreseeable future by a badly aligned AI. It doesn't really matter, from a truth-seeking perspective, whether the Rationalists are a load of misogynists; you can't psychoanalyse your way to the truth, as we discussed earlier. But it is easier to dismiss their concerns if you think they are.

Part Eight

Doing Good Better

Chapter 38

The Effective Altruists

It's impossible to talk about the Rationalists without mentioning their conjoined twin, the Effective Altruism movement. They're so intertwined that I have a bad habit of using the terms synonymously, but they are in fact distinct.

Effective Altruism is based on the idea that we should do *the most good we can* with our resources. If we give to charity, we should give to those charities that do the most good. If we want to dedicate our careers to improving the world, then we should think carefully (and use numbers) to establish *how* we should spend our careers.

The spiritual godfather of the movement is Peter Singer. He's an Australian moral philosopher, based at Princeton, who in 1972 published an essay entitled 'Famine, Affluence and Morality'.[1] Singer is a utilitarian, one of the most influential of the twentieth century. His argument in 1972 turned on two deceptively simple points: one, 'Suffering and death from lack of food, shelter and medical care are bad'; and two, 'If it is in our power to prevent something very bad from happening, without thereby sacrificing anything morally significant, we ought, morally, to do it.' He says, by way of illustration, 'If I am walking past a shallow pond and see a child drowning in it, I ought to wade in and pull the child out. This will mean getting my clothes muddy, but this is insignificant, while the death of the child would presumably be a very bad thing.'

That all seems pretty uncontroversial, certainly to me. But it has profound and unsettling implications. He was writing during the Bengal famine of 1971, at a time when 9 million people were refugees. He pointed out that Britain and Australia, two of the more generous countries in terms of aid, had given (in Britain's case) roughly one-thirtieth of the money it had spent on the Concorde project; Australia, one-twelfth of the amount it had spent on the new Sydney Opera House. That money could have saved many tens of thousands of lives. 'It makes no moral difference whether the person I can help is a neighbor's child ten yards from me or a Bengali whose name I shall never know, ten thousand miles away', he wrote. A life is a life, and we should save the most we can. It's not simply that giving money to aid charities is a nice thing to do, an over-and-above-the-call-of-duty moral bonus – 'supererogatory', in the language of moral philosophy – it's that it is a duty. We should all, in wealthy, developed countries, give some non-negligible percentage of our earnings, either as charitable donations or as foreign aid via tax, to improving the lot of people in developing countries. How much? He doesn't know, exactly, but 'It should be clear that we would have to give away enough to ensure that the consumer society, dependent as it is on people spending on trivia rather than giving to famine relief, would slow down and perhaps disappear entirely.'

Thirty-eight years after Singer wrote his essay, a small group of academics in Oxford, led by Toby Ord and William MacAskill, a professor of moral philosophy (at one point, the youngest associate professor in the world; he's still only 32), took Singer's ideas and tried to turn them into something both more practical and more precise. In 2009 they founded a charity called Giving What We Can, dedicated to finding the most effective charities in the world and helping people donate to them. It also encouraged people to pledge to donate 10 per cent of their income to those charities.

In 2011, MacAskill founded another charity, 80,000 Hours, which helps people to decide what career they should embark on in order to make the most positive impact on the world. For instance, while it might make intuitive sense, for a bright young person who wants to make a difference, to become a doctor at Médecins Sans Frontières, it might be more effective for her to go into some high-paying but less obviously worthy career, and then donate a large part of her salary to charity. (They cite an example of a man who chose to earn $250,000 a year as a software engineer instead of taking a $65,000 job as the CEO of a non-profit; he then donated $125,000 a year to charity, theoretically, at least, funding two CEOs.[2])

Over in the US, two young men had had a similar idea. In 2007, Holden Karnofsky and Elie Hassenfeld, Ivy League grads who had gone on to work in hedge funds, wanted to find out the best places to donate some of the moderately large amounts of money they were suddenly making. 'I think I just had this background assumption that it would be like buying a printer,' Holden told me. 'I'd go on some website and get the best printer for the least money.' But such a website did not appear to exist. 'I was like, "What am I trying to do? I'm trying to help people. How can I help people with a fixed amount of dollars that I'm going to give? It was just too hard to figure out." I started calling charities themselves and asking them: "Hey, do you have some metrics? Do you have some data on how many people you've helped?" Elie and I had a little club of eight people, all calling charities.'

Slowly they realised that the numbers were not easy to get hold of, and that the problem was quite big. 'At first it was annoying: it's holiday time and I'm spending all my time on the phone with charities trying to understand their numbers. Then I started to find it interesting. We saw a huge gap, a problem in the world, and we thought we could fix it if we worked really hard at it. So we left our jobs, raised money from our co-workers, and started GiveWell [and, later, OpenPhil]. It was a super-standard start-up story in a sense. I wanted something, I couldn't get it.'

You might, at this point, ask how you measure 'helping people'. How can you compare, say, the impact of an education programme to the impact of a medical programme? Is it fair or meaningful? 'Comparing across different types of charities is not only fair, but essential,' Will MacAskill told me. 'These comparisons allow us to see that some charities are much more cost-effective than others.' For instance, a charity that funds an extra year of education for people in the US or the UK might cost somewhere in the region of a few tens of thousands of dollars per person it helps. For that amount of money, you could provide thousands of insecticide-impregnated bed-nets to children in sub-Saharan Africa, which would prevent hundreds of cases of malaria and save several lives.

However much you value education, it's pretty clear that saving the lives of children is more valuable than a year of education for a Westerner.

This is a pattern that repeats. 'Charities that help people in developing countries tend to be much more cost-effective than charities that help people in developed countries,' says MacAskill. 'I call this the "100× multiplier": as someone living in a rich country, you should expect to be able to do at least 100 times as much to benefit other people as you can to benefit yourself or people in your community.' Both GiveWell and Giving What We Can recommend the Against Malaria Foundation, the charity which gives bed-nets to families in poor, malaria-prone regions. They also both suggest the Deworm the World Initiative, which pays for cheap deworming treatments which appear to have a significant impact on children's futures, and GiveDirectly, which makes small payments straight to poor people in developing countries via mobile phones. Other charities they both recommend include the Schistosomiasis Control Initiative, No Lean Season, Helen Keller International, the Malaria Consortium and the END Fund. What all these charities have in common is that they operate in poor countries; most of them also provide cheap medicine for easily treatable diseases. The Giving What We Can website offers this reasoning: 'Suppose we want to help people suffering from blindness. In a developed country this would usually involve paying to train a guide dog and its new owner, which costs around $40,000. In the developing world there are more than a million people suffering from trachoma-induced blindness and poor vision which could be helped by a safe eye operation, costing only about $100 and preventing 1–30 years of blindness and another 1–30 years of low vision . . . For the same amount of money as training one guide dog to help one person, we could instead prevent 400–12,000 years of blindness.'[3]

There are edge cases, and difficult things to compare: is extending a life by five years worth more or less than curing a child's blindness? But as it stands, most of us give to charity in spectacularly un-evidence-driven ways, sponsoring a friend's half-marathon or signing something a charity mugger hands us.

'Donors in developed countries who give in a non-evidence-based manner typically support charities that focus on helping people in their own communities,' says MacAskill. 'These charities can be expected to do one-hundredth or less of the good that charities focusing on helping people in poor countries do.' MacAskill is right: I was startled to learn, as I wrote this, that only 6 per cent of American charitable donations go to 'international affairs', which is to say 94 per cent of American charitable donations stay within American borders.[4] The idea of the Effective Altruism movement is that we are not, generally, dealing in subtle distinctions: if you want to do the most good with your money, rather than just purchase warm feelings, then some charities are very obviously better than others.

The links between the Rationalists and the Effective Altruists go back pretty much to the beginning. Ord and MacAskill met Nick Bostrom at Oxford in 2003, and Ord says: 'I was heavily influenced by Nick in my work on existential risk. I'm pretty sure [the Effective Altruism movement] wouldn't have had such a strong strand on existential risk if I hadn't been influenced by Nick.' It's not that one inspired the other, says Ord, 'the story is mostly one of mutual influence of people exploring related ideas and gaining from their interactions'.

Certainly, the LessWrong Rationalists provide a large proportion of Effective Altruism's support. In 2014, 31 per cent of survey respondents said that they had first heard of the movement through LessWrong;[5] by 2017, that figure had dropped to 15 per cent, presumably partly because LessWrong had shrunk while Effective Altruism had grown, but a further 7 per cent had heard of it through Slate Star Codex.[6] And a large fraction of LessWrongers are Effective Altruists: according to the 2016 LessWrong diaspora survey, 20 per cent of respondents identified as Effective Altruists, and 22 per cent had made 'donations they otherwise wouldn't' because of Effective Altruism.[7] Two other OpenPhil employees, Helen Toner and Ajeya Cotra – whom you'll meet shortly – told me that they'd either come to Effective Altruism through Less-Wrong or found the two at the same time. One Effective Altruist blog, 'The Unit of Caring', is named after a Yudkowsky blog post

('Money: The Unit of Caring'). Scott Alexander has written repeatedly about Effective Altruism, including an extremely powerful blog post[8] backing the Giving What We Can 'pledge' to donate 10 per cent of your income to effective charities.

Effective Altruists are also a similar sort of people to Rationalists. They are nerdy, on the whole. They are open to new experiences (I don't know how many of the Effective Altruists are polyamorous, but it's not uncommon, in both the Oxford and Bay Area sets). They go for numbers over feelings, even when the numbers lead them into weird areas, which they frequently do. For instance, one very good point that some Effective Altruists make is that if we are worried about alleviating suffering, then presumably animal suffering matters to some extent too. (You could declare by fiat that it doesn't, but it's hard to make a principled case for it. Besides, that would mean you're OK with the arbitrary torture of animals for no reason, which most of us are not.)

But if animal suffering matters, even a fairly small amount, then suddenly, eating chicken is an absolute moral catastrophe. Around 50 billion chickens are raised each year worldwide, the large majority of them for meat.[9] They live on average for seven weeks each, so that's about 6 billion chicken-years of life a year. Even if you only care about chickens' suffering-per-hour 1 per cent as much as you care about that of humans (a very conservative figure: Buck Shlegeris, the young Effective Altruist and Rationalist we met earlier, told me he rated it about a quarter, and Peter Singer told me that wasn't ridiculous), you should care as much about the world's chickens as you do about *the entire population of the United Kingdom*. And that's before you get to the point that chickens very probably have worse lives than the average Briton.

(Scott Alexander points out that you can improve the situation considerably by eating beef.[10] Imagine that you get a sixth of your calories from chicken, about 125,000 calories a year. A chicken provides about 3,000 calories. So your 125,000 calories translate to 42 chickens, bred and kept in unpleasant circumstances and then slaughtered. By contrast you get about 405,000 calories from a cow, so your 125,000 calories translates to 0.3 cows: just by switching, you reduce the number of animals bred for your food by 93 per cent. Cows live longer before slaughter, but not *that* much longer, and tend to have happier lives anyway. Unfortunately, cows are far worse for global warming, so really I ought to give up both and go vegetarian.)

But we can all get behind the idea that factory farming is a real problem and realise that we have a moral responsibility for it. Effective Altruism gets much weirder than that, in its niche areas. There are, for instance, Effective Altruists who worry about the suffering of wild animals. They make the point that there are probably between 100 million and 1 billion 'bugs' (insects, arachnids and so on) per human in the world, and that it only takes a very, very tiny level of moral value per insect to make their suffering outweigh that of all humans.[11] I can't fault the numbers, but they take me beyond the level of weirdness to which I am prepared to go. (And it gets weirder still. Another group has wondered about whether consciousness, and therefore suffering, is a fundamental quality of the universe – whether quarks and electrons are capable of suffering. There are quite a lot of quarks and electrons, so if you start worrying about that, you end up in a very odd place indeed.[12])

I shouldn't overplay the 'weirdness' element. Most Effective Altruism comes down to the eminently sensible idea that if you want the shiny pound in your pocket to do the most good it can, it's better to donate it to the Against Malaria Foundation than it is to your local donkey sanctuary (or even a cancer-research charity). If I keep going on about how weird they are, there's a risk that I'll make you think they're some strange species of humans. 'It's otherising,' says Karnofsky. 'The initial reaction to these topics is that it's wacky. But I think it's a really interesting and under-appreciated set of issues. If you other-ise them, you create this idea that you need a certain kind of unusual psychology to be interested in this stuff, and I don't think that's true.' The same goes if we paint them as super-altruistic: 'If you portray Effective Altruists as saints, as people who give away every penny they have, feel bad every time they buy a sandwich because they could have given it to charity, people read it and they go, "That person is weird. That person is not me. I could never relate to them, and now I'm not interested."'

But, again, a lot of the ideas are extremely straightforward.

That said, one key area that OpenPhil in particular is worried about, and which is definitely weird on the face of it, is existential risk. OpenPhil was set up by GiveWell and Good Ventures, a philanthropic foundation established by Dustin Moskowitz, one of Facebook's founders, and his wife Cari Tuna, although it is now separate from both those organisations. Where GiveWell concentrates on small, repeatable, reliable (in most cases) interventions, such as bed-nets or cash transfers, OpenPhil aims at big, high-risk-high-reward projects. One area it focuses on is US government policy – criminal justice reform, immigration policy and so on. It also works on animal-welfare issues; when I was in San Francisco talking to Karnofsky and his team, their head of comms, Mike Levine, took me out to lunch at a local burger bar, so he could gently probe me to make sure I wasn't going to write an entire book mocking the nerds who think Skynet is going to take over the world (he is endearingly protective of his nerds). We ate the Impossible Burger, a plant-based meat burger, which was produced partly with funding from OpenPhil,[13] the idea being to reduce the requirement for beef and thus improve animal welfare and reduce greenhouse-gas emissions. I can report that it tastes very much like a burger, indeed a perfectly nice burger, and appears to be taking off; it is sold at hundreds of locations in the US and one of its rivals has, since I went over there, arrived in Britain. (Specifically, and unsurprisingly, Dalston in east London.[14])

But the OpenPhil work that is most relevant to this book is its focus on global catastrophic risks, and especially AI.

Chapter 39

EA and AI

Ajeya Cotra is a research analyst for OpenPhil. She and her colleague Helen Toner, an Australian, both work on AI risk specifically, and both say that it is hard to explain to people what you do for a living. 'When I was in high school, I discovered GiveWell and Effective Altruism, and I also discovered LessWrong and the Rationalist community,' Cotra tells me. At first she was interested in global poverty reduction – 'I was trying to get my parents to donate to the Against Malaria Foundation' – and not focusing on AI. But at college in Berkeley she taught a seminar series on Effective Altruism. 'That was really when I had to force myself to think through all the arguments for [existential risk]. I started to transition to doing more work focused on our global catastrophic risks.' Her degree was in computer science, so she had some technical understanding of AI, which helped.

It has come with some social cost, she says. 'It's harder to explain, like to my parents, what I do. But I think you can understand fairly quickly why this is important. A lot of experts think that artificial intelligence might be arriving in the next 20 or 30 years. And human intelligence has had such a transformative, positive and negative, effect on our world.' There's no reason not to assume that even more powerful intelligence will have an even more transformative effect. 'It will touch on so many things we care about, like curing diseases, new forms of surveillance, wealth and poverty and inequality.'

Toner agrees that it's easier to say to people that you're working on anti-malarial bed-nets than it is to say you're trying to stop the world being destroyed by a rogue AI. She, too, started out worrying about global poverty and development. 'But after a long series of conversations, and learning more about stuff, and a year or two of being involved in the EA community, I started to come around to the view that maybe AI was the thing to be working on.'

There are three key elements that make a cause worth donating to, according to the tenets of Effective Altruism. One is its importance: the scale of the problem, how much better the world would be if the problem were solved. A second is tractability: how easy it is to solve those problems. And a third is neglectedness: if lots of people are already working on the problem, then the amount of good you can do on the margin is less. So malaria is an excellent target, because it has a huge impact (importance), is easily and cheaply prevented (with bed-nets), and yet receives far less global spend than diseases like cancer, which disproportionately affect those rich countries where people tend to live long enough to get them.

An obvious global catastrophic risk is climate change. It's hugely important, and while it's not easily tractable, there are certainly things that can be done. But climate change is a crowded space: there are lots of governmental and philanthropic organisations focusing on it, so the marginal value of the money OpenPhil puts in would be less.

Instead, the two risks that OpenPhil is focused on are pandemics, especially bioengineered pandemics, and AI. 'I've just gone back and forth enough times on which of those I consider a bigger risk,' Holden Karnofsky told me. 'If you asked me how [humans could go extinct], those are definitely my top two guesses by some margin and then a little bit down will be things like nuclear weapons.'

And, as we discussed in Chapter 2, talking about existential risks, going extinct *matters*. Remember that Nick Bostrom thinks that the number of human-like lives that could be lived is something obscene like 10^{58}. He could be wrong by three dozen orders of magnitude and we would still be talking about vastly more humans than have ever lived. If their lives have any non-negligible moral value (which is not, it should be said, an uncontroversial claim), then those potential lives need to affect our moral judgements significantly. Even if we don't buy that, merely a small chance of killing everyone alive is a vital issue: for instance, the expected population of the Earth by mid-century is about 10 billion. Something that has a 3 per cent chance of killing all those people is equivalent to something that will definitely kill 300 million. According to 80,000 Hours, that's 'more deaths than we can expect over the next century due to the diseases of poverty, like malaria'.[1]

If you buy into the idea that future lives matter, then you need to think about the things that are most likely to cause human extinction, not just those that will probably kill a large number of us. As we discussed a few chapters ago, it seems that bio-risk and AI are the best-value bets, and because this is a book about AI, I'm going to talk about that.

AI risk, says Holden Karnofsky, is extremely important (because of the risk of extinction), highly neglected (there are only a few places, such as MIRI and FHI, which are working on it), and reasonably tractable. So it represents an 'outstanding philanthropic opportunity'. He hasn't always thought this way: one of the most-read articles on LessWrong was a long piece by him about why he thought that MIRI was a bad subject for philanthropic investment. In a subsequent explanation of why his position changed, he said that he had previously thought that 'by the time transformative AI is developed, the important approaches to AI will be so different from today's that any technical work done today will have a very low likelihood of being relevant'.[2] But now, he told me, 'We think there's a non-trivial, by which we mean at least 10 per cent, chance of transformative AI in the next 20 years.'

That's part of what makes it tractable. The AI landscape has changed spectacularly in the last few years: things that were cutting-edge specialist technology, like voice and facial recognition software, in the early 2000s are now running on novelty apps on your phone. DeepMind's AlphaGo is a beautiful and, it must be said, slightly unnerving demonstration of how machine learning can create superhuman intelligence, in a more general (if still narrow) sense than we are used to. It's a testament to how quickly new stuff becomes normal that we aren't more amazed by it all.

But at the moment, the key thing that OpenPhil is doing isn't so much related to the technical work – although they do support that. They're excited about what Toner calls 'field-building'. She gives the example of the field of geriatric medicine. 'In the 1980s, people realised that elderly people need a totally different style of healthcare from young people. Young people tend to come in with one thing that needs to be fixed and then you send them home, whereas with elderly people there's a whole bunch of interacting conditions and maybe they need to stay a long time.' They also realised that the baby-boomer generation, at that point in its thirties and forties, was going to get old, which meant a very large population of elderly people. So the John A. Hartford Foundation decided to concentrate almost all its resources on building up the field of geriatric medicine, training doctors, funding research, building new centres. 'And by the time the [baby boomers were] elderly, geriatrics was a totally normal medical field,' says Toner.

That's roughly what OpenPhil is trying to do now. '[We've] been talking to high-profile, very skilled top machine-learning researchers, and saying, "Would you be interested in working on safety, or in having some grad students work on safety? What are the areas that you would be interested in?"' Toner says. In 2017 they gave \$3 million in grants to machine-learning departments at Berkeley and Stanford, about the same amount to MIRI (despite, a year earlier, having expressed 'strong reservations about MIRI's research'[3] while granting them \$500,000), and a much larger \$30 million to OpenAI, the non-profit research organisation founded by Elon Musk.[4]

The aim, says Karnofsky, is for their work to do the sort of field-building that the Hartford Foundation did, even though worrying about transformative and possibly disastrous AI might sound 'wacky' now. 'To an outsider who has never encountered the issues before, given the low-level of buy-in from wider society, the natural, initial reaction to these topics is going to be, "What is this? I've never heard of this. This is different from what I normally think about, this is wacky."' But in time, as the field becomes more mature and well known, he wants people to think more along the lines of, 'This is an issue that should matter to me. It's got a lot in common with climate change. It could be far off, but we don't *know* it's far off, and it's worth worrying about now because it could be a huge deal.' He hopes this book might make a few more people think like that.

I don't think *many* people will disagree with the idea that a pound spent trying to prevent malaria deaths will do more 'good', under most definitions of that word, than a pound given to support Harvard University. But you might remember what David Gerard said, when we were talking about whether LessWrong was a cult: 'clearly the most cost-effective initiative possible for all of humanity is donating to fight the prospect of unfriendly artificial intelligence, and oh look, there just happens to be a charity for that precise purpose right here! WHAT ARE THE ODDS.'⁵

The criticism that others have made is not that LessWrong is chiselling cash, but that the Rationalists, and the Effective Altruist movement, are heavily made up of nerdy, STEM-inclined computer-science grads. Dylan Matthews, writing for Vox in 2016, put it like this: 'In the beginning, EA was mostly about fighting global poverty. Now it's becoming more and more about funding computer science research to forestall an artificial intelligence-provoked apocalypse. At the risk of overgeneralising, the computer science majors have convinced each other that the best way to save the world is to do computer science research.'⁶ (It reminds me of a story that the Rationalist and Effective Altruist Ben Kuhn told on his blog. He went to an Effective Altruism summit with his partner, who was new to EA; she asked an attendee what sort of people were there. "'Oh, all different kinds!'" he replied. "'Mathematicians, and economists, and philosophers, and computer scientists . . ." It didn't seem to occur to the fellow that these were all basically the same kind of person,'⁷ Kuhn writes.)

This was, pretty much, the argument that Caroline Fiennes put to me. Caroline is the director of Giving Evidence, a group which encourages and enables giving based on sound evidence. She has known GiveWell and others in the Effective Altruism movement for a long time – she is on the board of The Life You Can Save, Singer's charity – and doesn't want to belittle what they do, but she is worried about this sort of uniformity of thought. 'People gravitate to stuff they understand,' she says. 'Maybe these people have gravitated towards this issue because they feel comfortable and competent on it.' Elon Musk, Peter Thiel, Dustin Moskowitz are the big-name funders behind it, and perhaps it isn't surprising that these software tycoons all think that good software is needed to save the world. And, as various people have muttered to me during the writing of this book, they seem to be less vocal about the sort of problems – privacy, surveillance, filter bubbles, fake news, algorithmic bias – that their software is creating. It's fun to sound noble and far-sighted, warning of the dangers of a technical problem still to come; it's less fun to address the criticisms people are making of what you and your peers are doing right now.

Fiennes notes that GiveWell and OpenPhil don't mention a lot of major areas of possible philanthropy. 'I don't think GiveWell's list of recommended charities really reflects global priorities,' she says, 'and it doesn't reflect the lists of other experts who do serious cause prioritisation, such as the 50 Breakthroughs report, or the Copenhagen Consensus. There's nothing about war, nothing about climate change. Only one thing, very recently added, about hunger, only one about water and sanitation. We have a couple of billion people who don't have access to food and clean water and toilets; it's weird that those issues aren't further up the list.' There's also nothing about global governance and the rule of law, she adds, 'which for me, and for George Soros I would observe, would seem like a reasonable thing, if you had a chunk of money'. GiveWell will do the 'things that are very certain, very repeatable, very proven – we know the impact of the next bed-net, know the costs, know the benefits,' while OpenPhil is willing to talk about really long-shot things like bioterrorism and superintelligent AI. 'But there's all this big pile of stuff in the middle [like promoting the rule of law], which seems obviously important. I'm not saying AI isn't important,' Fiennes says, 'but the amount I hear about AI out of them, compared to the nothing at all I hear about global governance and climate change and war, seems totally out of whack.'

It's not completely fair to say that the two organisations – they're separate now, although they still share the same swanky office space in downtown San Francisco – ignore these issues completely. Mike Levine, OpenPhil's head of comms, agreed that climate change, for instance, is not as neglected or as tractable as some other, similarly important, issues. 'But I'd quibble with the idea that we don't focus on it. We've funded the most important, neglected and tractable climate-change opportunities we've seen, including geoengineering research and governance and the especially tractable opportunity around the Montreal Protocol [to reduce the use of hydrofluorocarbons, a powerful greenhouse gas]. We've put millions of grant dollars into climate change – more than we have into our land-use reform, macroeconomic stabilisation, or immigration policy focus areas. We see climate change as a huge issue that warrants action from philanthropists, and we expect to do more.' But it's certainly true that these mid-level issues get less *attention*, in the press and public sphere, than OpenPhil's concentration on AI.

That may or may not be OpenPhil's (and the wider Effective Altruism community's) fault, depending on your point of view. You could say that they simply need to focus on what they think are the most effective targets for their philanthropy, and let other people worry about the look of the thing.

But it's also true that 'how weird it looks' affects how effective it is. Will MacAskill told me that he thinks there is a 'pretty solid' case for focusing on AI safety, for those people who are 'prepared to consider more speculative lines of evidence . . . though I should point out that some of the arguments for concluding that one should focus on existential risk reduction exclusively make a number of controversial philosophical assumptions', such as that future lives are of comparable value to current ones. But there is a risk, he said, that 'an excessive focus on these speculative causes runs the risk of undermining the Effective Altruism movement'. If people look at Effective Altruism and see a bunch of people worrying about what seems to them to be some sci-fi stuff, they might then not donate to the less speculative things – the bed-nets and the cash transfers – which they mentally bucket together. It could be, in Karnofsky's word, 'other-ising'.

There's a concept in the Rationalsphere called 'weirdness points'. It's the idea that society will let you be only *so* weird before it stops taking you seriously: you only have a certain number of weirdness points to spend, and so you should spend them on things that you really care about. That's why, says Mike Story, Rationalists don't talk all that much about their polyamory – 'more than just not evangelise, they keep quiet about it. Scott [Alexander] is pretty open about it, but generally they think if we seem normal it's better for our ideas.' If you spend your weirdness points on polyamory, you don't have them left to spend on Effective Altruism or the importance of Bayes' theorem.

And, it must be said, AI safety spends a *lot* of weirdness points. If you think it's the most valuable thing by miles, then it is worth spending those points on. If you don't, you could reasonably argue that it's just making the Effective Altruism movement look weird and making it harder to get bed-nets to children in sub-Saharan Africa.

So it comes down, really, to whether we can trust the thinking of OpenPhil and similar organisations on all of this. I'll raise a couple of possible points. One was an objection that Dylan Matthews of Vox had, in the piece I mentioned previously. He raises the possibility of 'Pascal's mugging', which we discussed before. People at the Effective Altruism conference he attended gave him the standard expected-value argument: 'Infinitesimally increasing the odds that 10^{52} people in the future exist saves way more lives than poverty reduction ever could.' But Matthews argues that the key is what we mean by 'infinitesimally': 'Maybe giving $1,000 to the Machine Intelligence Research Institute will reduce the probability of AI killing us all by $[10^{-17}]$,' he writes. 'Or maybe it'll make it only cut the odds by $[10^{-66}]$. If the latter's true, it's not a smart donation; if you multiply the odds by 10^{52}, you've saved an expected $[10^{-13}]$ lives, which is pretty miserable. But if the former's true, it's a brilliant donation, and you've saved an expected $[10^{34}]$ lives.'

On the face of it, that sounds a pretty fair argument – 'those probability values are just made up,' says Matthews. 'I don't have any faith that we understand these risks with enough precision to tell if an AI-risk charity can cut our odds of doom by $[10^{-17}]$ or by only $[10^{-66}]$. And yet for the argument to work, you need to be able to make those kinds of distinctions.'

I'm going to hand over to Scott Alexander to explain why he (and I) think that argument doesn't work. In short, it's that while 10^{-66} and 10^{-17} both sound like similar spectacularly tiny numbers, they are absolutely not. The number 10^{-17} is the sort of number you might actually have to use at some point. There are about 2,522,880,000 seconds in an 80-year human life, and about 7 billion humans, so during your lifetime about 1.7×10^{19} seconds will be lived by humans. That means you'd expect, in your lifetime, roughly 100 things to happen that only happen once in every 10^{17} seconds. (When you read about ridiculously unlikely things happening – someone finding out that their next-door neighbour is actually their separated-at-birth sibling, or whatever – remember that.)

But 10^{-66} is *not like that*. You will *never* read about something that only happens once every 10^{-66} seconds. Here's Scott: 'The per-second probability of getting sucked into the air by a tornado is 10^{-12}; that of being struck by a meteorite 10^{-16}; that of being blown up by a terrorist 10^{-15}. The chance of the next election being Sanders vs Trump is 10^{-4}, and the chance of an election ending in an electoral tie about 10^{-2}. The chance of winning the Powerball is 10^{-8} so winning it twice in a row is 10^{-16}. Chain all of those together, and you get 10^{-65}.' (He was writing in 2015, hence the 'Sanders vs Trump' reference.)

What this means is that, if Matthews' 10^{-66} guess is right, then then the likelihood of your $1,000 donation to MIRI helping avert an AI apocalypse is 10 times less than the likelihood of someone getting simultaneously sucked up into the air by a tornado, hit by a meteorite and blown up by a terrorist, on the same day as winning the lottery for the second week in a row, and Trump and Sanders tying the electoral college. If Bostrom's numbers are anywhere near accurate – the 10^{58} human lives that we talked about early on in this book – and if you accept that future lives matter as much as current ones, then even a vanishingly tiny chance that an AI disaster might happen is (from a utilitarian point of view) hugely important. You have to start making some seriously weird assumptions to get it down to a negligible figure.

The other possibility, of course, is that you argue that future lives don't matter as much as current lives. That's a massive and ongoing bunfight in moral philosophy, and, as MacAskill said, there's no clear answer on it. When I spoke to Peter Singer, he pointed out that there's a difference between *future lives* and *possible lives*. 'It's highly probable that there will be humans living on this planet in 100 years, and their lives are going to be worse off because of climate change. We shouldn't discount those lives just because they'll only exist in 100 years.' But when we're dealing with lives in the very distant future, where it is far from certain that they'll exist at all, he said, 'we are not talking about the sufferings of beings who will exist in future. We are talking about the non-existence of many generations of future beings. It remains a controversial and disputed question of philosophy. A number of very good philosophers have worked on it; Derek Parfit worked on it for most of his philosophical career and was unable to really achieve a satisfying conclusion.'

Obviously, we're not going to resolve it here. But, Singer points out, we don't *really* need to. We're uncertain; these lives might be exactly as important as modern lives, or they might not be important at all. But if we're *really* uncertain, then we can't be sure that they have zero value. 'Because of that uncertainty, we should give at least some weight to them. If we say, "No, they don't count" – well, we have to acknowledge that might be the wrong thing to say.'

And then we're back to the 'Bostrom big number' thing. 'If there are so many of them as Bostrom argues,' says Singer, 'even if you discount them by 99 per cent or 99.9 per cent, the numbers are so big that they still carry very great weight. It's a tough question, but yes, we should give some weight to the interest of merely possible beings. We should regard the extinction of a species as a worse event than the deaths of all the people who will be killed at that time.'

Of course, even if we agree that the thronging masses of possible future people really do have some moral weight, that preventing extinction is therefore valuable, and that AI is one of the most realistic ways in which extinction could happen and is worth funding, that doesn't mean that OpenPhil is correct in assessing that MIRI (or FHI, or OpenAI) in particular is worth backing. There are strong criticisms of MIRI: that its output of scientific papers is rather less than that of a single grad student, for instance. (There is a counter-argument that that's not a fair criticism, because it's trying to build a new field rather than get on the publication treadmill in established journals, but there's a lot of back-and-forth about it.)

And there is cause to be sceptical of some of Effective Altruist reasoning. GiveWell's top charities include several that promote deworming. This comes largely from a study, published in 2004,[8] which found that mass deworming – giving all the children at a school deworming tablets, not just those with worms – improved health, school performance and school attendance; and not just the children at that school, but at schools miles away, through stopping the worms from spreading. Most amazingly, it seemed to dramatically increase those children's earnings later in life. It was an extremely cheap intervention with, apparently, enormous results.

But later studies looked at the data and found that it was flawed in some quite serious ways: technical but important statistical errors which severely undermined its credibility. (Not, I hasten to add, through any deliberate misbehaviour on the part of its au-thors, who very nobly gave up their data for it to be checked; just mistakes.) The Cochrane Collaboration, which does huge meta-analyses of all the studies on a topic, has since looked at mass deworming three times;[9] it found no effect on school performance or attendance, and no good evidence of an effect on various other health measures. 'Going up against the Cochrane Collaboration is big bananas,' says Caroline Fiennes. 'And Paul Garner, who does their parasitology, has been doing this for a hundred years. I'm amazed that GiveWell isn't more alarmed that people who have thought long and hard about this – studied deworming since before they were born, in some cases – disagree with them.' It's not just that deworming might not do any good – it's perfectly con-ceivable that it might do harm. 'I spoke to a parasitologist who's studied deworming in Africa for decades,' Fiennes continues, 'who said that mass deworming has never been adequately tested for increasing drug resistance, so mass deworming may just jack up resistance and actually be harmful.'

GiveWell is aware of these criticisms. Catherine Hollander, a GiveWell research analyst, told me they'd run the data, and looked at the ways it wasn't robust, but felt it was still a worthwhile gamble because it was so cheap and had the potential to be so effective. 'Even when you discount for the possibility that that effect doesn't exist significantly, you still end up with something that's the most cost-effective thing that we recommend,' she said. 'We discount extremely heavily for all of this uncertainty. But even with that huge discount, Deworm the World was 10 times as cost-effective as cash transfers, our next most cost-effective programme.'

I, obviously, cannot reasonably assess whether deworming is a good bet; nor can I do a more reliable job than Bostrom or OpenPhil in assessing the numbers behind AI risk. But for what it's worth, I think Fiennes' criticisms are worth taking on board. They *should* make you wary of going along too happily with what these Effective Altruism organisations recommend; and, as we've discussed before, if something seems completely weird, but the numbers check out, you should pay some attention to the feeling that it's weird. The Dylan Matthews Vox piece refers to Effective Altruists who seemed to think that literally all philanthropy should go towards preventing human extinction – which would be a pretty terrifying situation.

But as it stands, that's not the case. The total worldwide spend on AI risk reduction is probably less than $50 million, which may seem like a lot, but is only one-eighth of (to pick a big charity at semi-random) Greenpeace's reported annual revenues in 2014.[10] (OpenAI's funders 'have committed $1 billion', but the organisation expects 'to only spend a tiny fraction of that in the next few years'.[11]) It is not that AI risk is crowding out all other charity; it's not even monopolising Effective Altruism. It accounted for about 30 per cent of total OpenPhil grants in 2017, but OpenPhil is only one of several organisations. And Peter Singer, who is very probably the best-known moral philosopher alive today and the absolute godfather of the field of efficient charity, is (broadly, and with caveats) in favour of spending money on AI risk: 'I definitely think it's worth looking at,' he told me. 'It's definitely worth spending something on it. Even if the probability is very low, given how devastating it could be, it's worth trying to put some effort into reducing it even further.'

Part Nine

The Base Rate of the Apocalypse

Chapter 40

What are they doing to stop the AI apocalypse?

Cast your mind back to several chapters ago, when we were talking about why AI is dangerous. We mentioned that there are several organisations – MIRI is the one I've spent most time talking about, but also Bostrom's FHI, Max Tegmark's Future of Life Institute in Cambridge (Massachusetts), the Centre for the Study of Existential Risk in Cambridge (UK), and Elon Musk's OpenAI are the obvious other ones – which are dedicated, at least in part, to making AI less dangerous; some of them are definitely Rationalist groups, others are just broadly aligned. But I haven't really told you, yet, what they're actually *doing*.

It is, of course, an open question whether there's anything they can do, at this stage at least. We're still years, probably decades and possibly centuries away from AGI. Murray Shanahan of DeepMind isn't convinced that the field is mature enough for the work we do now to have significant effect when AGI does happen: 'If you want my personal view,' he says, 'I think that it's probably a bit premature to be very confident that the work we're doing now on this issue is going to be relevant.' He's not ruling it out, but he's wary. 'We really don't know what AGI is going to look like, if and when we figure out how to make it.' He thinks that the MIRI safety work, which looked at 'logical problems related to insuring that self-modifying systems could preserve their reward functions, and things like that', was 'fascinating stuff and quite mathematically demanding'. But, he says, MIRI's work revolves around an assumption that any AGI will be an extremely logical Bayesian-probability-theory engine, and it's possible to imagine an AI that is nothing like that at all. 'Suppose you could solve AI with some massive evolutionary process, and you just evolved your AI, or if you just had some enormous deep network with some fabulous amount of computation and back-propagation, you might not have any way of applying [MIRI's results] to the thing you've built, because it wouldn't operate in a sufficiently logical way.'

When he mentioned this, I asked if AlphaGo was an example of the sort of program he was talking about. It was given a goal – become amazingly good at Go – and a reward mechanism, and then it was sent to go and play against itself billions of times until it got good. And the thing that came out was, indeed, amazingly good at Go, but no one who built it really knows *why*. The data and learning mechanisms went into a black box, sloshed around for a bit, and came out with the thing they wanted.

'Well, exactly,' Shanahan says. 'And you might still be in a position to design a reward function that you know isn't going to have unintended consequences and perverse instantiations – the paperclip maximiser, these side effects that involve existential risk. But I'm involved in the sharp end of building these things, and making them increasingly powerful, and I just have difficulty extrapolating from where we are now to AGI, and being confident about what it's going to look like.'

Obviously enough, Rob Bensinger of MIRI is more optimistic that the work they're doing will be useful. 'I think there are lots of particular problems which you can work on today. There's no particular reason to think we've grabbed all the low-hanging fruit yet. There are presumably lots of problems that you can't see in advance, and you need to work with the system and learn about them later, but there's plenty that's knowably relevant right now.' One aspect they're working on, in fact, is avoiding exactly the sort of 'impenetrable black-box' scenario that Shanahan is talking about. 'We want AGI systems, when they're built, to be well understood by their developers, and to minimise the assumptions we have to make about your system in order to be confident they're safe,' he said. That could mean making a system which just has really simple outputs – for example, something that just comes up with mathematical proofs and theorems, and doesn't do anything else.

But that's pretty limited: 'The reason you build an AGI in the first place [is that] you want to get useful work out of it. Cure disease, help people, whatever. Set theory proofs aren't the kind of things that are really going to make a difference in the world.'

So instead you want to be able to see *how* the AGI is doing its work. 'Not in every detail, but in broad strokes,' said Bensinger. '"What is this part here? What does this system do?"' He drew a diagram on a nearby whiteboard as he spoke, boxes with arrows to other boxes, like a flowchart. 'You can look at its internals and see the kind of optimisation that went into the final product, and confirm that it doesn't have any bad convergent instrumental goals and does have the properties you *do* want. We assume that AGI will be more complex than current systems, so it'll get complicated. But MIRI's view is that the important thing is to empower the developer of the first really powerful AGI system to know what they're doing, to be able to design a really modular, clean system where you can tell a story about why that system would have good effects. It wouldn't be a perfect story, there will always be some uncertainty. But if you can't even tell a story about why you should be confident – not just, "We don't know why we *shouldn't* be confident" . . .' He didn't finish the thought, but the implication left hanging was 'that wouldn't be great'.

And you can't get that sort of story out of a black-box system, he said. 'If you just get the outputs and say, "Well, it seems to act safely, nothing bad has happened so far", that's not the sort of story we think you can get an outcome out of.' So the sort of evolved AI that Shanahan mentioned would make MIRI very nervous indeed. As, in fact, would AlphaGo. 'I'd say you're in a bad state if you kind of understand your system, but the understanding is too many degrees removed from actual cognitive work that's happening. Things like AlphaGo, where you can describe it at a high level – it's using Monte Carlo tree search, it has these value networks, you can say this stuff at a certain level of abstraction – and then you get a really good Go player. But you're not really describing the reasoning it's doing to get into good board positions.' You're describing how it built itself, but not what it's doing. I think it's fair to say that if DeepMind use AlphaGo as the basis for their first AGI, MIRI would be deeply unnerved.

But 'Make it so we can see how it works' is quite a high-level description of the method for creating a safe AI. Remember, way back when we were discussing *why* AI is dangerous, we talked about how seemingly innocuous goals can have weird outcomes, 'perverse instantiations': the paperclip maximiser was one example, the *Fantasia* broom another. Is anyone doing anything more specific to try to reduce the likelihood that, the first time an AGI is turned on, it will turn the solar system into computing hardware to become even better at chess?

The answer is yes. There are a few things. One of the key ones, in fact, is clearly delineating what the AI safety problems actually are. Probably the most famous paper on AI safety is 'Concrete Problems in AI Safety', by a team from Google, OpenAI, Berkeley and Stanford.[1] It discusses, for instance, the problem of defining negative side effects: if you give an AI a goal (such as 'fill the cauldron'), how do you let your AI know that there are things to avoid, without explicitly listing every single thing that could go wrong? An example of how you could do so is by saying that the robot can only change the environment it's in by a certain amount – limiting or budgeting its impact on the world – although how you define that is a new and perhaps equally tricky problem.

Another problem the paper identified was 'reward hacking', an AI finding a shortcut to achieving its utility function without doing what its makers want; 'ending cancer by nuking humanity' would be a dramatic example, although the more prosaic one they mention in the paper is of a cleaning robot being told to stop when it can't see any more rubbish, so it just turns off its cameras. The paper suggests that a separate agent designed solely to judge whether the rewards the first agent receives are earned might be a good way around it.

As well as simply defining the problems, there are specific efforts to find solutions. Holden Karnofsky, the OpenPhil founder, got quite excited talking to me about a paper called 'Deep Reinforcement Learning from Human Preferences',[2] by, among other people, Shane Legg, one of the founders of Deep-Mind, and Paul Christiano of OpenAI.

The idea is that for some goals, a simple definition isn't much use. Karnofsky showed me a video of a simulated environment called MuJoCo, a sort of toy world with broadly realistic physics that programmers use to test how a robot might move. First, he showed me three 'robots' in that world which had been given a task, 'Learn to walk.' They all made reasonable progress; one was a snake-thing which wriggled along, one was a uniped which hopped, the third bipedal and vaguely humanoid.

'They've all been given the goal of trying to cover a lot of distance in a short amount of time,' said Karnofsky. 'They have these limbs and they have these joints. What they do is move around at random, and over time they figure out that certain movements make more progress, and over time they learn to walk. You can run the same algorithm on a bunch of different-looking robots and they'll all learn to walk in their own special way.' That's because 'walking', or at least 'making progress', is nice and easy to define mathematically. 'You have an X-coordinate,' said Karnofsky, 'and the further you get from it per unit of time, the better you're doing. It's measurable.'

But it relies on that score providing good feedback about what you *actually want*. In general (although not always, as we saw earlier when we were talking about evolved AIs), 'how good is this robot at walking?' and 'distance travelled from starting point in given time' align extremely well. That won't always be the case in real-world AIs, though. Karnofsky imagines a more advanced AI. 'Something that not only is able to move joints, but is able to send e-mails and make business decisions and do all kinds of things in the world. You can imagine if you had one of those AIs you can say, "Hey, can you maximise the amount of money in this bank account."' It's a nice, simple, mathematically defined task, although any of us can see how it might go appallingly wrong.

Tasks that might *not* go immediately wrong are harder to define. 'You could ask it, "Hey, can you stop other AIs from doing bad things on their way to maximising money", and that would be a great [example of a] bad task, because I myself don't know what I mean by that,' said Karnofsky. '"Maximise the money" is a well-defined goal. "Keep us safe" or "Help humanity thrive" or

"Make the world more peaceful" are fuzzily defined goals. I don't even know what I mean by them.'

What the Christiano paper does is try to see whether we can get AIs *today* to learn to perform tasks that we can't define well. 'That's where the human feedback comes in,' Karnofsky said. 'The idea is that the AI will try random movements and then an actual human will look at two videos and say, "That one. That one is more like what I have in mind."'

In the MuJoCo paper, they try this on a backflip. It turns out that it's quite hard to define a backflip in a mathematically precise way, and all the robots that tried to learn how to do one with a preset goal ended up doing weird jerky movements that absolutely were not backflips. But with a human saying, 'That random movement looks a bit more backflip-y' they evolved a pretty impressive-looking backflip quite quickly – it took 900 iterations and less than an hour. Another task it learned was to play a racing video game, called Enduro. But not just play it to win. 'They got it to keep pace with other cars,' Karnofsky said. 'The video game does not reward that. There's no score for it; it's just the human was able to look at it and say, "This is what I wanted."'

This is early, toy stuff. And it's not perfect: in one task, the AI was supposed to learn to pick up a box with its manipulators, but instead learned to put a manipulator between the box and the camera so it looked to the operator as though it had done so. But you can see how it could be extrapolated into a more powerful machine with more complex goals: 'Rid the world of cancer', but don't do it by killing all the humans, that sort of thing.

Another step, Karnofsky added, would simply be to get governments and tech companies to sign treaties saying they'll submit any AGI designs to outside scrutiny before switching them on. It wouldn't be iron-clad, because firms might simply lie, 'but it's substantially better than nothing. There's a good chance that when the first transformative AI comes it'll be a massive project.' If there are 1,000 people working on it, it'll be almost impossible to keep it secret, so the treaties might be quite effective.

I asked Nick Bostrom what he thought the most promising avenue was. 'Broadly speaking,' he replied, 'some way that involves leveraging the AI's intelligence to infer and learn about human values and preferences. If it's superintelligent, it should be able to figure out what we want and mean, just as I can figure out a lot about what you want, from asking you about it, and looking at what you choose and so forth.'

More specifically, he pointed to another idea from Paul Christiano, on the topic of 'capability amplification'.[3] 'Instead of trying to create this AI that has a utility function that captures everything we care about,' said Bostrom, 'we have some agent that, at each point in time, has some set of available actions, and chooses the one it thinks we would most approve of it taking, in some myopic way.' It doesn't try to leap ahead and think what will happen in six months' time, it just chooses between A and B, and then between B1 and B2, and so on. 'If you have to have a human overlooking every single step the AI takes, you don't get much oomph out of it, so you'd have to come up with clever ideas for how to bootstrap a more limited, but safe, system for doing a larger portion of what we'd want the superintelligence to do.'

But, he said, it's still early days. 'There are lots of other ideas, and it might be that the best ideas haven't even been articulated yet. Or maybe we already have a solution, but we can't be confident yet that it would work because there are some parts we don't yet understand. There's just a lot of uncertainty about the difficulty of the problem.'

I should, also, point out that there's a lot of scepticism about just how much good MIRI in particular is doing. They have several researchers, but publish very few papers which are rarely cited. That might be because they're trying to build a field, as discussed before, but it does make it hard to be sure how important their work is at this stage. Toby Ord of FHI told me that he was somewhat sceptical of their research agenda, which he felt was a bit too focused on distant-future ideas: 'I'm more excited about work that's more continuous with current work in AI,' he said. 'Stuff that's only one step away from what the AI researchers are working on. There's more chance that you can get them to come across and help you on the project.'

Of course, technical papers about how to keep AI safe aren't the only measure of how successful the Rationalists are being. Remember that when the young Eliezer Yudkowsky started thinking about this stuff in about 2000, he was pretty much the first person to do so. People had discussed the 'singularity', and I.J. Good had predicted an 'intelligence explosion', but the specific problem that an AI might not 'go rogue' or 'turn evil' or 'achieve self-awareness' but simply do exactly what you told it to do, and still go terribly, terribly wrong, seems to have sprung up with Bostrom and Yudkowsky on the SL4 website. It was, and I hope this isn't too rude, the random musings of a load of young, crankish men on the early internet.

Now it's the subject of whole departments at several major universities, and discussed by major leading intellectuals: Martin Rees, the late Stephen Hawking, Bill Gates. Google's DeepMind is explicitly worried about it; co-founders Shane Legg and Demis Hassabis both take it seriously.

'It's been amazingly quick progress,' commented Ajeya Cotra, of OpenPhil. 'I think the discourse around technical safety of these in 2014, compared to now, feels like different worlds.' Bostrom's book *Superintelligence* was a turning point, she said. It was a *New York Times* bestseller, and brought serious academic heft to the field. 'In 2014, we had a small group of outsider futurists trying to convince the AI community that this is a risk to be taken seriously. A lot of researchers didn't, because they heard a garbled message through the media which sounded kind of fearmongery. Then *Superintelligence* came out. It was less that a bunch of people read the arguments and were totally convinced than that it was a serious academic making a thorough case. People felt it merited a response, and when sceptics put forward that response, their points didn't always add up.'

This roughly matches Holden Karnofsky's position. 'As of 2012, it was kind of a niche community issue, and it was very hard to find anyone with mainstream credentials who would even acknowledge it,' he said. He also pointed to the publication of *Superintelligence* as a turning point, and a major AI safety conference in Puerto Rico in 2015, organised by Max Tegmark's Future of Life department at MIT.

'They had an open letter saying AI has major risks, which was signed by a lot of people. I don't think all of them were signing off on [AI risk as envisioned by Yudkowsky et al.]. But it certainly became a more mainstream idea to talk about the idea that AI is risky.' Now, he said, 'if you look at the top labs, you look at Deep-Mind, OpenAI, Google Brain – many of the top AI labs are doing something that shows they're serious about this kind of issue.' 'Concrete Problems in AI Safety' is a collaboration between those three groups, he pointed out. 'It's a paper where they have safety in the title and it has major researchers from all these three labs and it was put out under the Google PR machine. It's much more mainstream than it used to be. It seems hard to deny that.'

All this has meant that, from a field-building point of view, Yudkowsky et al. have been extraordinarily successful. 'There are well-known, well-respected machine-learning researchers involved now,' said Helen Toner, Ajeya Cotra's OpenPhil colleague. 'How to build this field is to make it something young people feel comfortable going into – not just starry-eyed Rationalists who think they want to go and save the world, but talented young machine-learning researchers who are searching for an area to specialise in. I heard of, I think at Berkeley, a student-supervisor pair, who were each interested in AI safety, and each of them thought, "Oh I can't tell my supervisor/my student about that because they'll think it's silly."' 'Concrete Problems in AI Safety' and various OpenPhil grants to major machine-learning research groups have started breaking that dynamic down, she commented. 'We're really trying to fix those dynamics, and I think it's helped.'

'It's incredible!' said Paul Crowley, beaming. 'Honestly, when I got involved I thought there's no way we'll get anyone to take this seriously – people will just think we're crazy. There's no way we can be more than some crazy fringe thing that a few people talk about.'

Chapter 41

The internal double crux

Right at the beginning of the book, I mentioned what Paul Crowley said to me: that he doesn't expect my children to die of old age. What I didn't say, back then, was whether I took him seriously. Do I actually think that is more likely than not?

Rather than simply answer that question, I want to talk you through an experience I had in Berkeley. Soon after I met Paul, I spoke to Anna Salamon of CFAR, and told her what Paul had said. It threw me, I said. The thing that I really like about the Rationalist community, I told her, is the idea of people trusting the numbers and the reasoning, and not throwing out the conclusion those numbers and reasons lead them to, even if it's shocking or bizarre. But now I found myself in a strange situation, where I was pretty happy with all the different steps in the reasoning – I can see why value alignment might be hard, and why an AI could be amazingly intelligent but still do stupid things, and I don't think it's crazy to think it'll happen in my children's lifetimes – but I found putting them all together, and agreeing with that profoundly unnerving conclusion, difficult.

To her enormous credit, Anna didn't, as most people would, simply dismiss my concerns about the central thesis of the movement of which she's a crucial part; instead, she pointed out that to have such concerns is not a stupid thing to do. *She* thinks the technological singularity, and all the disaster or utopia that entails, probably will happen this century. But when I made that remark about instinctively rejecting the conclusion despite agreeing with the steps to get there, she shrugged. 'Yeeaaaaaah,' she said. 'But that's not always wrong, is it? When I was first taking algebra class, and somebody showed me the standard proof that 2 = 1 – which involves secretly dividing by zero – it felt to me like every single step was valid.' When you're dealing with complex things like this, if you get weird answers, it might be that you've input the steps wrong. 'I think the thing to do in such a case isn't to reject one thing or the other, but to really stay with the question. Having now stayed with the question for a while, it seems to me that the argument for there being substantial risk from AI is really quite strong.'

Then she asked: 'Do you believe it? That your children will not die of old age? That *you* might not die of old age?' I said I didn't know. I couldn't see the divide-by-zero equivalent – I still can't. But nor could I, on an intuitive level, feel that the conclusion was right. I still assume that I'll grow old and die, as my grandfather had recently. And while I don't like to think about it, I assume that my children will too. There was a tension between these two parts of my brain.

'So,' Anna said, 'would you like to try something? We have a funny CFAR technique called the internal double crux.' The *standard* 'double crux' is a method Rationalists use for examining why two people disagree. 'It's about how to figure out what cruxes a disagreement you're having with someone rest on. The crux of the argument is the thing that, if you knock it over, their conclusion falls down, and they have a different conclusion.'

The example given on LessWrong is an argument between two people about school uniforms.[1] Person A thinks schoolchildren should wear uniforms; person B thinks they shouldn't. To find the crux, you look at what those beliefs entail, what the more specific implications of them are. So person A might think that school uniforms reduce bullying, by making it less obvious which children are rich and which are poor; person B might think that's ridiculous. But if you could show that school uniforms do reduce bullying, by some given amount, then person B would change her mind on the uniforms question; likewise, if you could show that they don't, then person A would change his mind. The technique involves slowly bringing the conversation away from top-level, shouty arguments and towards detailed, specific disagreements.

This is pretty useful and sensible, I think. But the internal double crux is a bit more strange. It's for disagreements with yourself – for instance, if part of you agrees with the argument that says, 'Your children won't die of old age', and part of you thinks that just sounds too crazy to be believed. 'It's a pretty weird thing,' Anna warned me. 'It involves going into your head, and sometimes you find things there that sound a little crazy. But I could walk you through it.' So, a few days later, I went back to the CFAR offices.

There's a little story that Anna tells, a sort of parable, about a little girl at school who does some creative writing, and at the end of it the teacher reads it through, and says, 'Look, you misspelled "ocean".' 'No I didn't,' says the kid, and the teacher replies: 'I'm sorry, but you did. It's counter-intuitive, but it's a "c" not "s-h".' The child, increasingly angry, repeats, 'No, I didn't,' the teacher, 'No, I'm sorry but you did.' 'No, I didn't.' 'I realise it sometimes hurts to face the truth, but you really did.' And then the child runs off to the cupboard and bursts into tears, saying, 'I did *not* misspell the word. I *can too* be a writer.' Anna calls this a bucket error. The 'I can spell the word "ocean"' fact went into the same bucket as the 'I can be a writer when I grow up' fact, and when the one was proved false, the child assumed, on some pre-conscious level, that they both were.

More broadly, there's often a strange lack of communication between our verbal, reasoning selves and some deeper part of us. The feeling that you've forgotten something at the supermarket, but you don't know what it is – is it broccoli? No. Is it rice? No, but I do need rice; is it avocado? It's avocado! The knowledge is there on some level, and the feeling when your conscious mind latches onto the right answer is one of almost physical relief. Or something gives us a near-physical sense of 'yuck' or 'yum' when we see or hear it, without exactly telling us why, and unless we give that sensation a few moments we might not know the reason; the word 'slack' gives me this sort of 'yuck' sensation, and for years I didn't know why, until I remembered an *Onion* article which used it in an astonishingly vulgar phrase about female genitalia which apparently had stuck with me ever since.

According to the way in which CFAR and the Rationalists model the brain (and it is just a model; Anna repeatedly stressed that she wasn't claiming that this is really how the brain works, just that it was useful and effective), sometimes, when we have internal conflicts about an issue, a little alarm, a beeping noise, goes off when we take on information; our brain is telling us that the information conflicts with something somewhere, and although we may not consciously know why, it imbues that piece of information with a little sense of 'yuck' and makes it harder for us to accept it. This is, I think, what Eliezer Yudkowsky describes as 'noticing your confusion'. When I said to Anna that each of the steps made sense, but that I couldn't accept the conclusion, she wondered if that was what was going on. The 'internal double crux' technique is a method Rationalists use in this sort of situation to help them establish what their own objection is, to find their difficulty with some conclusion.

I am aware that this all sounds a bit mystical and self-helpy. It's not. It was a strange sensation, certainly, but it was extremely common-sensical and unspectacular, although it did make me understand how talking therapy (which I've never tried) could be a powerful tool: just the experience of *talking very deeply* about some mental experience you've had is quite profound. Anyway, the basic idea was simple. Anna asked me: 'What's the first thing that comes into your head when you think the phrase, "Your children won't die of old age"?'

'The first thing that pops up, obviously,' I told her, 'is I vaguely assume my children will die in the way we all do. My grandfather died recently; my parents are in their sixties; I'm almost 37 now. You see the paths of a human's life each time; all lives follow roughly the same path. They have different toys – iPhones instead of colour TVs instead of whatever – but the fundamental shape of a human's life is roughly the same. But the other thing that popped up is a sense of "I don't know how I can argue with it", because I do accept that there's a solid chance that AGI will arrive in the next 100 years. I accept there's a very high likelihood that if it does happen then it will transform human life in dramatic ways – up to and including an end to people dying of old age, whether it's because we're all killed by drones with kinetic weapons, or uploaded into the cloud, or whatever. I also accept that my children will probably live that long, because they're middle-class, well-off kids from a Western country. All these things add up to a very heavily non-zero chance that my children will not die of old age, but, they don't square with my bucolic image of what humans do. They get older, they have kids, they have grandkids and they die, and that's the shape of the life. Those are the two fundamental things that came up, and they don't square easily.'

So Anna asked me to look at the two sides in turn.

The more sceptical bit of me said this. 'There's some quite big bit of me which looks at all this stuff and thinks, *you're talking about immortality, and the end of the world, and all these things that people have prophesied since for ever.* Every generation thinks this one's the last one. There's not a generation in history that hasn't thought exactly that, and with every single one it's been ridiculous. We look back at the Heaven's Gate people, or Christian prophets in the first century, or the Anabaptists in Reformation Münster, all thinking it's going to happen any day now; we've been through this story a million times before, and, every time, it doesn't happen. The lesson we should draw from all this is that the predictions of the doom, or ascension, of mankind tend not to come through.'

OK, said Anna. Take that, and offer it up to the other bit of you and see what it says; let it, in her words, 'fully and generously acknowledge' all the parts of the statement that seem to be true, giving it room to breathe.

And the bit of me that accepts the steps of the argument had this to say. 'All right, there really have been a lot of predictions, and from the point of view of the people making the predictions, it must have felt as if the logic of all the parts fitted together, or they wouldn't have made the predictions. I assume these people weren't idiots. They must have had reasons to believe that the end times were on them. It would be very hard for someone inside that chain of logic to step outside and say, "Yes, but there have been lots of predictions before, and they were wrong." You have to use the base rate as a real piece of evidence in your thinking about these things. Otherwise you're doing no better than the people who made all those predictions before. You have to look at the base rate, and the base rate of predictions of the apocalypse coming through is zero.' This is the 'outside view', which we talked about a few chapters ago. The internal logic of something can be compelling, but you have to look at how other, similar things have fared in the past.

Then the worried bit was allowed to reply: 'But this is also an argument against taking steps against climate change. You're saying the world always carries on as it always has, but, actually, we are changing the world. And we know that life on Earth is wildly different from how it was 100 years ago, and that was wildly different from how it was 300 years before that, but actually that was *less* different than 1,000 years before *that*. The pace of change of human life is gathering.' It's not ridiculous to think that it could be even more unrecognisable in 100 more years. And my worried self had a wider point. 'Every prediction of the doom of mankind will be wrong apart from one. That's implicit in the logic of it. If you keep saying the sun won't rise tomorrow, you'll be wrong every time until you're right. Induction can only work so many times; eventually, something must, by its nature, break it.'

So, the sceptical bit of me had to take that and acknowledge it. 'OK. I admit that a lot of the arguments that one could make to say that climate change is nothing to worry about are echoes of those made by sceptics who say, "Humanity has been on the planet for 100,000 years and we haven't caused the sea levels to rise yet." That doesn't mean that we won't soon, and we have impacts that are far greater now than they were even 50 or 100 years ago. And it's true that you can't have a base rate of doomsday predictions that is other than zero. If there are better reasons to believe in this one than the Anabaptists' prophecies, then we ought to take it on its own merits. Some of the things on which the Anabaptists based their arguments I would absolutely reject as being without foundation, whereas this prediction is based on things that I would not.'

Then that part of me was allowed to say: 'But you can track concentrations of carbon dioxide in the atmosphere, and make simple mathematical models to predict what will happen to things like temperature and sea-level rise, using equations that have been around since the early 1900s, and see that those predictions roughly match real-world data. I'm not sure of the extent to which that is comparable to artificial intelligence. And there is certainly more uniformity among climate scientists that climate change is imminent and dangerous than there is uniformity among AI researchers that AI is imminent and dangerous.' (I promise this is all real. I've tidied the quotes up a bit, but there's something about talking to Rationalists that makes you use phrases like 'heavily non-zero probability' and 'the base rate of predictions of the apocalypse'.)

It was at this point that the conversation, if that's the right word, took a slightly odd turn. It was still my sceptical side's turn to speak, and it had this to say: 'I can picture a world in 50 or 100 years that my children live in, which has different coastlines and higher risk of storms and, if I'm brutally honest about it, famines in parts of the world that I don't go to. I could imagine my Western children in their Western world living lives that are not vastly different to mine, in which most of the suffering of the world is hidden away, and the lives of well-off Westerners largely continue and my kids have jobs. My daughter is a doctor and my son is a journalist, whatever. Whereas if the AI stuff really does happen, that's not the future they have. They have a future of either being destroyed to make way for paperclip-manufacturing, or being uploaded into some transhuman life, or kept as pets. Things that are just not recognisable to me. I can understand from Bostrom's arguments that an intelligence explosion would completely transform the world; it's pointless speculating what a superintelligence would do with the world, in the same way it would be stupid for a gorilla to wonder how humanity would change the world.'

And I realised on some level that this was what the instinctive 'yuck' was when I thought about the arguments for AI risk. 'I feel that parents should be able to advise their children,' I said. 'Anything involving AGI happening in their lifetimes – I can't advise my children on that future. I can't tell them how best to live their lives because I don't know what their lives will look like, or even if they'll be recognisable as human lives.'

I then paused, as instructed by Anna, and eventually boiled it down. 'I'm scared for my children.' And at this point I apologised, because I found that I was crying. 'I cry at about half the workshops I do,' said Anna, kindly. 'Often during the course of these funny exercises.'

Chapter 42

Life, the universe and everything

I don't want to claim that the fact that I cried in a Californian office in the autumn of 2017 means that AI is going to kill us all and destroy the universe. I was alone in a strange country, 5,000 miles from my children, tired and jet-lagged and generally in quite an emotionally vulnerable situation. I don't cry very often, but it's not all that surprising that I did at this point. It was, though, quite powerful. I have tried to recapture the feeling of that moment and largely failed, although I felt an echo of it as I listened to the audio while writing this. It felt, emotionally, real that I didn't want to think about the implications because the implications were so terrifying.

So do I believe in a paperclip apocalypse? Let's think about this probabilistically.

I do believe that AGI could happen quite soon. It's far from clear when, but if a large number of AI researchers think it could well occur in the next 50 years and is near-certain in the next 100, then I don't know why I would disagree. Bostrom's survey, which seems to be the best we have, said that AI researchers, on average, think there's a 90 per cent chance that AGI will arrive by 2075.[1] For what little it's worth, I (like Bostrom) put more weight on the idea that it will never arrive, or will arrive in some immensely distant future, for no better reason than that AI, like fusion power, has been 30 years away for the last 50 years. Let's say I think it's 80 per cent likely that AGI will arrive at some point in the next 90 to 100 years, the likely lifespan of my children.

The next question is whether I think that is likely to lead to the sort of spectacularly terrible outcome that MIRI and other people fear. Going back to Bostrom's survey, 18 per cent of respondents believe that AGI will lead to something 'extremely bad' (existential catastrophe), i.e. human extinction. Some people who work on AI risk reckon it's higher: Rob Bensinger described it as 'high-probability', for Nick Bostrom the 'default outcome' of an intelligence explosion is 'doom'.[2] But some AI researchers consider that ridiculous: to Toby Walsh, for instance, the basic premise of an intelligent thing destroying the world was silly, and not really in keeping with what 'intelligence' means. Having looked into this stuff for quite a long time, I think he's probably wrong, but he's an AI researcher and I'm not. That fits the survey's findings that a majority of AI researchers don't think it's the most likely outcome, though, so I'm still going take that 18 per cent at face value.

Shut up and multiply, as the Rationalists say. If you take that 80 per cent (the likelihood that AGI will arrive at some point in the next 90 to 100 years) and multiply it by 18 per cent, you get 14.4 per cent, or almost exactly a one-in-seven chance. I might be off by quite a distance in either direction, but it feels about right. (Toby Ord's estimate for the likelihood of humanity going extinct this century, from any cause, is about one in six. 'I think it's something in the order of Russian roulette,' he told me, and that's *after* taking into account the fact that people are trying to stop it.) Imagine I'm off by more than an order of magnitude, because I might be; imagine there's only a 1 per cent chance that I'm right. That's still more likely than my children dying in a car crash, a risk which I do not think it is silly to worry about at all. I'm extremely happy to spend time teaching them the Green Cross Code, and for the government to invest in traffic-calming measures and impose legal safety requirements on car manufacturers to reduce that less-than-1-per-cent risk still further.

They're a weird bunch, the Rationalists, with their polyamorousness and kink, abstruse jargon, living arrangements and behaviour. And they're politically daft as well: their openness to debate means that their places on the internet are full of pretty unpleasant people. So the things they care about, such as AI risk and effective altruism, are in danger of getting smeared by association. 'The singularity? Isn't that the thing that racist sex-cult website is into?'

But you can't psychoanalyse your way to the truth. They might be weird, but I don't think they're *wrong* in believing that AI risk, like pandemics and climate change, is something society should be taking steps to mitigate. There's a non-trivial chance that it will do terrible things, and there are, it seems to me, realistic ways of trying to reduce the chance of that happening.

I met a senior Rationalist briefly in California, and he was extremely wary of me; he refused to go on the record. He has a reputation for being one of the nicest guys you'll ever meet, but I found him a bit stand-offish, at least at first. And I think that was because he knew I was writing this book. He said he was worried that if too many people hear about AI risk, then it'll end up like IQ, the subject of endless angry political arguments that have little to do with the science, and that a gaggle of nerdy Californian white guys probably weren't the best advocates for it. This was a concern I heard from a few people.

I hope that doesn't happen. The Rationalists are an interesting bunch of people and the things they're doing seem worthwhile: not just the AI risk, although of course that's central; but also the idea of thinking about *how* we think, and how we argue; and of considering probabilities and likelihoods in ways other than black and white and yes and no. And, yes, they can be hard to like – Eliezer Yudkowsky, in particular, is a difficult, strange and abrasive man, though undeniably smart and arguably visionary – and there are plenty of unpleasant people in their internet circles.

But there's something noble about their endeavour. We are, or seem to be, increasingly bad at talking across disagreement: whether because of social media or political polarisation, it seems that people find it much harder to imagine that someone with different political views to their own might nonetheless be a decent person. There's something wonderful about a project dedicated to explaining why we are so often wrong, and to taking arguments and ideas seriously, and not rejecting the ones we don't like: creating a space for people to disagree in good faith.

And what they have achieved in terms of the AI debate is, I think, remarkable. They've taken the niche, practically dystopian-science-fiction idea of AI risk and made people take it seriously.

Mike Story pointed out that Donald Glover, the actor and rapper who stars in the TV show *Community*, uses Bostrom's ideas in one of his episodes, indicating how mainstream these ideas have become. Perhaps more relevantly, the White House under Obama published a report in 2016 into 'Preparing for the Future of Artificial Intelligence'[3] which drew heavily and obviously on Yudkowsky/Bostrom ideas: the Puerto Rico conference and 'Concrete Problems in AI Safety' are extensively referred to, and Bostrom's work is cited in the references.

The Rationalist community has changed enormously since its mid-2000s heyday. Eliezer Yudkowsky himself has largely withdrawn from the LessWrong stuff; he writes the occasional very long blog post on the MIRI website, and some more playful and/ or not-AI-related stuff on Facebook and Tumblr, but he's less engaged with the community-building stuff these days.

But the rest of them carry on. Holden's OpenPhil still pushes money towards AI safety; Ajeya is still trying to build the field, although Helen has moved on and is now working on machine learning at a Chinese university. Paul remains at Google, encrypting things, but worrying about the future of humanity in his spare time. Scott and Katja broke up, but went on holiday together by accident a few weeks later, which felt very on-brand; they didn't have a baby, and I don't know whether the experimental robot baby helped with any decisions. They're both still highly engaged in the Rationalist project: Scott as its most high-profile figurehead now that Yudkowsky has taken a back seat; Katja as a researcher at FHI and elsewhere. Anna continues to see lots of bright young nerds come through the CFAR doors, and she tries, with great sensitivity and sense, to direct them into careers that might help save the planet. Rob is still Yudkowsky's messenger on Earth. And Nick Bostrom, of course, is now a globally famous figure in certain niche circles, and is getting referenced in major NBC comedy programmes and White House policy documents.

Overall, they have sparked a remarkable change. They've made the idea of AI as an existential risk mainstream; sensible, grown-up people are talking about it, not just fringe nerds on an email list. From my point of view, that's a good thing. I don't think AI is definitely going to destroy humanity. But nor do I think it's so unlikely that we can ignore it. There is a small but non-negligible probability that, when we look back on this era in the future, we'll think that Eliezer Yudkowsky and Nick Bostrom – and the SL4 email list, and LessWrong.com – have saved the world. If Paul Crowley is right and my children don't die of old age, but in a *good* way – if they and humanity reach the stars, with the help of a friendly superintelligence – that might, just plausibly, be because of the Rationalists.

Acknowledgements

It's frankly weird that this book got written. I am not the sort of person who successfully writes books. I have, traditionally, been the sort of person who periodically says, 'One day I'll write a book,' while the people around me say, 'Righto, Tom, you said that five years ago.'

So, I can take only a fraction of the credit for the fact that it did, in fact, get written. A large amount of the rest should go to these various other people.

Will Francis, agent at Janklow & Nesbit, kept taking me out for nice lunches until a serviceable idea fell out of my head. Paul Murphy at Weidenfeld & Nicolson helpfully agreed to give me some money to write the book, and then made me take out some of the sillier jokes. Linden Lawson did a sterling job clearing up my waffly and repetitious prose.

I owe great thanks to Ajeya Cotra, Andrew Sabisky, Anna Salamon, Buck Shlegeris, Caroline Fiennes, Catherine Hollander, David Gerard, Diana Fleischman, Helen Toner, Holden Karnofsky, Katja Grace, Michael Story, Mike Levine, Murray Shanahan, Nick Bostrom, Peter Singer, Rob Bensinger, Robin Hanson, Scott Alexander, Toby Ord, Toby Walsh and everyone else who spoke to me. Plus grudging thanks to Eliezer Yudkowsky who did not, in fact, agree to talk to me, but did answer my irritating questions by email. Elizabeth Oldfield and Pete Etchells both looked over parts of the manuscript and reassured me that it was not total garbage.

Especial thanks to Paul Crowley, who basically introduced me to the whole concept, invited me out to California, was extremely nice to me while I was there, answered loads of questions, and then read the manuscript and helped me strip out lots of stupid mistakes.

And, of course, Alison and Andy, my parents, for everything;

Emma, my wife, for everything else; and Billy and Ada, for not coming upstairs and hammering on the keyboard with their sticky little paws too often.

Notes

Introduction: 'I don't expect your children to die of old age'

1. Elon Musk, Twitter, 3 August 2014 https://twitter.com/elonmusk/status/495759307346952192?lang=en
2. https://qz.com/698334/bill-gates-says-these-are-the-two-books-we-should-all-read-to-understand-ai/
3. Cambridge University press release, 19 October 2016 http://www.cam.ac.uk/research/news/the-best-or-worst-thing-to-happen-to-humanity-stephen-hawking-launches-centre-for-the-future-of
4. Nick Bostrom, *Superintelligence: Paths, Dangers, Strategies* (OUP, 2014), p. 222
5. https://en.wikipedia.org/wiki/2017_California_wildfires

1: Introducing the Rationalists

1. http://yudkowsky.net/obsolete/singularity.html
2. *Omni* magazine, January 1983
3. 'Raised in technophilia', LessWrong sequences, 17 September 2008 https://www.readthesequences.com/RaisedInTechnophilia
4. 'The magnitude of his own folly', LessWrong sequences, 30 September 2008 https://www.readthesequences.com/TheMagnitudeOfHisOwnFolly
5. Nick Bostrom, 'A History of Transhumanist Thought', *Journal of Evolution and Technology*, vol. 14, issue 1, 2005 https://nickbostrom.com/papers/history.pdf
6. Marie Jean Antoine Nicolas Caritat, Marquis de Condorcet, *Esquisse d'un tableau historique des progrès de l'esprit humain* (Masson et Fils, 1822)
7. Benjamin Franklin, letter to Jacques Dubourg, 1773, US government archives https://founders.archives.gov/documents/Franklin/01-20-02-0105
8. Julian Huxley, *Religion Without Revelation* (Harper Brothers, 1927)
9. Eliezer Yudkowsky, *My life so far*, August 2000 http://web.archive.org/web/20010205221413/http://sysopmind.com/eliezer.html#timeline_great
10. William Saletan, 'Among the Transhumanists', *Slate*, 4 June 2006 https://web.archive.org/web/20061231222833/http://www.slate.com/id/2142987/fr/rss/
11. Quoted in Bostrom, 'A History of Transhumanist Thought', p. 14
12. Alvin Toffler, *Future Shock* (Turtleback Books, 1970)

13. Eliezer Yudkowsky, 'Future shock levels', SL4 archives, 1999 http://sl4.org/shocklevels.html

14. Eliezer Yudkowsky, 'The plan to Singularity', 2000 http://yudkowsky.net/obsolete/plan.html

15. 'Re: the AI box experiment', SL4 archives, 2002 http://www.sl4.org/archive/0203/3141.html

16. Nick Bostrom, 'The simulation argument', SL4 archives, 2001 http://www.sl4.org/archive/0112/2380.html

17. History of LessWrong, https://wiki.lesswrong.com/wiki/History_of_Less_Wrong

18. Overcoming Bias: about http://www.overcomingbias.com/about

19. 'Fake fake utility functions', LessWrong sequences, 6 December 2007 http://lesswrong.com/lw/lp/fake_fake_utility_functions/

20. Riciessa, 'LessWrong analytics, February 2009 to January 2017', LessWrong, 2017 https://www.lesswrong.com/posts/SWNn53RryQgTzT7NQ/lesswrong-analytics-february-2009-to-january-2017

2: The cosmic endowment

1. 'Research priorities for robust and beneficial artificial intelligence: An open letter', Future of Life Institute https://futureoflife.org/ai-open-letter

2. Donald E. Brownlee, 'Planetary habitability on astronomical time scales', in Carolus J. Schrijver and George L. Siscoe, *Heliophysics: Evolving Solar Activity and the Climates of Space and Earth* (Cambridge University Press, 2010)

3. Nick Bostrom, 'Existential risk prevention as global priority', 2012 http://www.existential-risk.org/concept.pdf

4. Carl Haub, 'How many people have ever lived on Earth?', 2011 http://www.prb.org/Publications/Articles/2002/HowManyPeopleHaveEverLivedonEarth.aspx

5. Bostrom, *Superintelligence*, p. 101

6. Ibid., p. 102

7. Existential Risk FAQ, Future of Humanity Institute http://www.existential-risk.org/faq.html

8. Eliezer Yudkowsky, 'Pascal's mugging: Tiny probabilities of vast utilities', LessWrong, 2007 http://lesswrong.com/lw/kd/pascals_mugging_tiny_probabilities_of_vast/

9. Nick Bostrom, 'Pascal's mugging', 2009 https://nickbostrom.com/papers/pascal.pdf

10. Scott Alexander, 'Getting Eulered', 2014 http://slatestarcodex.com/2014/08/10/getting-eulered/

11. Scott Alexander, 'Stop adding zeroes', 2015 http://slatestarcodex.com/2015/08/12/stop-adding-zeroes/

3: Introducing AI

1. Stuart J. Russell and Peter Norvig, *Artificial Intelligence: A Modern Approach* (3rd edn; Pearson, 2010), p. 1
2. A.M. Turing, 'Computing machinery and intelligence', *Mind*, vol. 59, 1950, pp. 433–60
3. Russell and Norvig, *Artificial Intelligence*, p. 3
4. Luke Muehlhauser and Anna Salamon, 'Intelligence explosion: evidence and import', 2012 https://intelligence.org/files/IE-EI.pdf
5. Eliezer Yudkowsky, 'Expected creative surprises', LessWrong sequences, 2008 http://lesswrong.com/lw/v7/expected_creative_surprises/
6. Eliezer Yudkowsky, 'Belief in intelligence', LessWrong sequences, 2008 http://lesswrong.com/lw/v8/belief_in_intelligence/
7. Ibid.
8. Demis Hassabis et al., 'Mastering chess and shogi by self-play with a general reinforcement learning algorithm', Arxiv, 2017 https://arxiv.org/pdf/1712.01815.pdf
9. Russell and Norvig, *Artificial Intelligence*, p. 4
10. Nick Bostrom and Vincent C. Müller, 'Future progress in artificial intelligence: A survey of expert opinion', *Fundamental Issues of Artificial Intelligence*, 2016 https://nickbostrom.com/papers/survey.pdf
11. Nick Bostrom, 'How long before superintelligence?', *International Journal of Future Studies*, vol. 2 1998 https://nickbostrom.com/superintelligence.html

4: A history of AI

1. Alan Turing, 'On computable numbers, with an application to the *Entscheidungsproblem*', *Proceedings of the London Mathematical Society*, vol. s2-42, issue 1, 1 January 1937, pp. 230–265 https://doi.org/10.1112/plms/s2-42.1.230
2. J. McCarthy, M. Minsky, N. Rochester and C.E. Shannon, 'A proposal for the Dartmouth summer research project on artificial intelligence', 2 September 1956. Letter to the Rockefeller Foundation, retrieved from http://raysolomonoff.com/dartmouth/boxa/dart564props.pdf
3. I.J. Good, 'Speculations concerning the first ultraintelligent machine', *Advances in Computers*, vol. 6, 1965
4. Charles Krauthammer, 'Be afraid', *The Weekly Standard*, 26 May 1997 http://www.weeklystandard.com/be-afraid/article/9802#!
5. John McCarthy, quoted in David Elson, 'Artificial intelligence', *The Johns Hopkins Guide to Digital Media* (Johns Hopkins University Press, 15 April 2014)
6. A. Newell, J.C. Shaw and H. A. Simon, 'Chess-playing programs and the problem of complexity', *IBM Journal of Research and Development*, vol. 2(4), 1958, pp. 320–335

7. Wolfgang Ertel, *Introduction to Artificial Intelligence* (Springer, 1993), p. 109

5: When will it happen?

1. Eliezer Yudkowsky, 'There's no fire alarm for artificial general intelligence' https://intelligence.org/2017/10/13/fire-alarm/
2. Donald B. Holmes, *Wilbur's Story*, 1st edn (Lulu Enterprises, 2008), p. 91 https://books.google.co.uk/ books?id =ldxfLyNIk9wC&pg= PA91&dq=%22i+said +to+my+brother+orville%22&hl= en&sa=X&redir_ esc=y#v=onepage&q= %22i%20said%20to%20my%20brother %20 orville%22&f=false
3. Richard Phodes, *The Making of the Atomic Bomb* (Simon & Schuster, 2012), p. 280 https://books.google.com/books?id= aSgFMMNQ6G4C&pg= PA813&lpg=PA813&dq= weart+fermi&source=bl&ots=Jy1pBOUL10&sig= c9wK_yLHbXZS_GFIvoK3bgpmE58&hl= en&sa=X&ved= 0ahUKEwjNofKsisn WAhXGlFQKHbOSB1QQ6AEIKTAA#v= onepage&q=%22ten%20per%20cent%22&f=false
4. Bostrom and Müller, 'Future progress in artificial intelligence' https:// nickbostrom.com/papers/survey.pdf
5. K. Grace et al., 'When will AI exceed human performance? Evidence from AI experts', ArXiv https://arxiv.org/pdf/1705.08807.pdf?_sp=c803ec8d-9f8f-4843-a81e-3284733403a0.1500631875031
6. David McAllester, 'Friendly AI and the servant mission', Machine Thoughts blog, 2014 https://machinethoughts.wordpress.com/2014/08/10/ friendly-ai-and-the-servant-mission/
7. Luke Muelhauser, 'Eliezer Yudkowsky: Becoming a rationalist', Conversations from the Pale Blue Dot podcast, 2011 http:// commonsenseatheism.com/?p=12147
8. Toby Walsh, *Android Dreams* (Hurst & Company, 2017), p. 54

6: Existential risk

1. Eliezer Yudkowsky/MIRI, 'AI as a positive and negative factor in global risk', 2008 https://intelligence.org/files/AIPosNegFactor.pdf
2. The Giving What We Can pledge: https://www.givingwhatwecan.org/ pledge/
3. Nick Beckstead and Toby Ord, 'Managing risk, not avoiding it', *Managing Existential Risk from Emerging Technologies*, Annual Report of the Government Chief Scientific Adviser 2014, p. 116 https://www.fhi. ox.ac.uk/wp-content/uploads/Managing-existential-risks-from-Emerging-Technologies.pdf
4. A. Robock, et al., 'Multidecadal global cooling and unprecedented ozone loss following a regional nuclear conflict', *Earth's Future*, 2014 http:// onlinelibrary.wiley.com/doi/10.1002/2013EF000205/full

5. 'Soviets close to using A-bomb in 1962 crisis, forum is told', *Boston Globe*, 13 October 2002 http://www.latinamericanstudies.org/cold-war/sovietsbomb.htm

6. List of nuclear close calls, Wikipedia https://en.wikipedia.org/wiki/List_of_nuclear_close_calls

7. Yudkowsky, 'AI as a positive and negative factor in global risk' https://intelligence.org/files/AIPosNegFactor.pdf

7: The cryptographic rocket probe, and why you have to get it right first time

1. Nate Soares, 'Ensuring smarter-than-human intelligence has a positive outcome' 2017 https://intelligence.org/2017/04/12/ensuring/

2. Tom Chivers, 'The spaceship that took some of the greatest images of the solar system has died', BuzzFeed, September 2017 https://www.buzzfeed.com/tomchivers/cassini-death-spiral

8: Paperclips and Mickey Mouse

1. Nick Bostrom, 'Ethical issues in advanced artificial intelligence', 2003 https://nickbostrom.com/ethics/ai.html

2. http://www.decisionproblem.com/paperclips/index2.html

3. Soares, 'Ensuring smarter-than-human intelligence has a positive outcome' https://intelligence.org/2017/04/12/ensuring/

9: You can be intelligent, and still want to do stupid things

1. Bostrom, *Superintelligence*, p. 9

2. Nick Bostrom, 'The superintelligent will: motivation and instrumental rationality in advanced artificial agents', 2012 https://nickbostrom.com/superintelligentwill.pdf

3. Elezier Yudkowsky, 'Ghosts in the machine', 17 June 2008 https://www.readthesequences.com/GhostsInTheMachine

4. NCD Risk Factor Collaboration, 'Trends in adult body-mass index in 200 countries from 1975 to 2014: A pooled analysis of 1698 population-based measurement studies with 19.2 million participants', *The Lancet*, 2 April 2016 http://www.thelancet.com/journals/lancet/article/PIIS0140-6736(16)30054-X/fulltext

5. Yudkowsky, 'AI as a positive and negative factor in global risk', 2008 https://intelligence.org/files/AIPosNegFactor.pdf

6. S. Omohundro, 'The basic AI drives', 2008 https://selfawaresystems.files.wordpress.com/2008/01/ai_drives_final.pdf

7. Bostrom, 'The superintelligent will' https://nickbostrom.com/superintelligentwill.pdf

10: If you want to achieve your goals, not dying is a good start

1. Soares, 'Ensuring smarter-than-human intelligence has a positive outcome' https://intelligence.org/2017/04/12/ensuring/
2. Omohundro, 'The basic AI drives' https://selfawaresystems.files. wordpress.com/2008/01/ai_drives_final.pdf
3. Thucydides, *History of the Peloponnesian War*, trans. Richard Crawley (J.M. Dent & co., 1903), Chapter 1 https://ebooks.adelaide.edu.au/t/ thucydides/crawley/complete.html
4. Hans Morgenthau, *Politics Among Nations: The Struggle for Power and Peace* (McGraw-Hill Education, 1967), p. 64
5. Thomas Hobbes, *Leviathan*, 1651 (Andrew Crooke, 1st edn), Chapter 13
6. Nikita Khrushchev, 'Telegram From the Embassy in the Soviet Union to the Department of State', 2 October 1962. From *Foreign Relations of the United States*, 1961–63, Volume VI, Kennedy-Khrushchev Exchanges, US Department of State Office of the Historian, ed. Charles S. Sampson, United States Government Printing Office 1966 https://history.state.gov/ historicaldocuments/frus1961-63v06/d65
7. Bostrom, 'The superintelligent will' https://nickbostrom.com/ superintelligentwill.pdf
8. Soares, 'Ensuring smarter-than-human intelligence has a positive outcome' https://intelligence.org/2017/04/12/ensuring/
9. Maureen Dowd, 'Elon Musk's billion-dollar crusade to stop the AI apocalypse', *Vanity Fair*, April 2017 https://www.vanityfair.com/ news/2017/03/elon-musk-billion-dollar-crusade-to-stop-ai-space-x
10. https://wiki.lesswrong.com/wiki/Roko's_basilisk
11. http://rationalwiki.org/wiki/Roko%27s_basilisk/Original_ post#Comments_.28117.29
12. https://xkcd.com/1450/
13. https://www.reddit.com/r/xkcd/comments/2myg86/xkcd_1450_aibox_ experiment/cm8vn6e/
14. David Auerbach, 'Roko's Basilisk, the single most terrifying thought experiment of all time', *Slate*, 17 July 2014 http://www.slate.com/articles/ technology/bitwise/2014/07/roko_s_basilisk_the_most_terrifying_thought_ experiment_of_all_time.single.html
15. Dylan Love, 'Just reading about this thought experiment could ruin your life', *Business Insider*, 6 August 2014 http://www.businessinsider.com/what-is- rokos-basilisk-2014-8?IR=T
16. 2016 LessWrong diaspora survey results http://www.jdpressman. com/public/lwsurvey2016/Survey_554193_LessWrong_Diaspora_2016_ Survey%282%29.pdf
17. Scott Alexander, 'Noisy poll results and reptilian Muslim climatologists from Mars', 2013 http://slatestarcodex.com/2013/04/12/noisy-poll-results- and-reptilian-muslim-climatologists-from-mars/

11: If I stop caring about chess, that won't help me win any chess games, now will it?

1. Omohundro, 'The basic AI drives' https://selfawaresystems.files. wordpress.com/2008/01/ai_drives_final.pdf

12: The brief window of being human-level

1. Vernor Vinge, 'Signs of the Singularity', 2008 http://www.collier.sts.vt.edu/engl4874/pdfs/vinge_2008.pdf
2. Demis Hassabis, et al.,'Mastering the game of Go with deep neural networks and tree search', *Nature*, January 2016 https://www.nature.com/articles/nature16961
3. Miles Brundage, 'AlphaGo and AI progress', February 2016 http://www.milesbrundage.com/blog-posts/alphago-and-ai-progress
4. Eliezer Yudkowsky, 'My Childhood Role Model', 2008 https://www.readthesequences.com/MyChildhoodRoleModel

13: Getting better all the time

1. Bostrom, *Superintelligence*, p. 2
2. Robin Hanson, 'Economics of the Singularity', 1 June 2008 https://spectrum.ieee.org/robotics/robotics-software/economics-of-the-singularity
3. I.J. Good, 'Speculations concerning the first ultraintelligent machine', *Advances in Computers*, vol. 6, 1966, pp. 31–88 http://commonsenseatheism.com/wp-content/uploads/2011/01/Good-Speculations-Concerning-the-First-UltraIntelligent-Machine.pdf
4. Bostrom, 'The superintelligent will' https://nickbostrom.com/superintelligentwill.pdf
5. Ibid.
6. Yudkowsky, 'AI as a positive and negative factor in global risk', 2008 https://intelligence.org/files/AIPosNegFactor.pdf

14: 'FOOOOOM'

1. Bostrom, *Superintelligence*, p. 65
2. Ibid., p. 65
3. Ibid., p. 70
4. Ibid., p. 68
5. Eliezer Yudkowsky, 'Hard takeoff', LessWrong, 2008 http://lesswrong.com/lw/wf/hard_takeoff/
6. Luke Meuhlhauser and Anna Salamon, *Intelligence Explosion: Evidence and Import* (MIRI, 2012) https://intelligence.org/files/IE-EI.pdf

15: But can't we just keep it in a box?

1. Scott Alexander, 'No physical substrate, no problem', 2015 http://
slatestarcodex.com/2015/04/07/no-physical-substrate-no-problem/
2. Dowd, 'Elon Musk's billion-dollar crusade to stop the AI apocalypse'
https://www.vanityfair.com/news/2017/03/elon-musk-billion-dollar-crusade-
to-stop-ai-space-x
3. Bostrom, *Superintelligence*, p. 129
4. 'The "AI box" experiment', SL4 archives, 2002 http://www.sl4.org/
archive/0203/3132.html
5. Eliezer Yudkowsky, 'Shut up and do the impossible!', LessWrong
sequences, 2008 https://www.lesswrong.com/posts/nCvvhFBaayaXyuBiD/
shut-up-and-do-the-impossible
6. Bostrom, 'Risks and mitigation strategies for Oracle AI', 2010 https://
www.fhi.ox.ac.uk/wp-content/uploads/Risks-and-Mitigation-Strategies-for-
Oracle-AI.pdf

16: Dreamed of in your philosophy

1. Bostrom and Müller, 'Future progress in artificial intelligence' https://
nickbostrom.com/papers/survey.pdf
2. Scott Alexander, 'AI researchers on AI risk', 2015 http://slatestarcodex.
com/2015/05/22/ai-researchers-on-ai-risk/
3. Ibid.

17: 'It's like 100 per cent confident this is an ostrich'

1. Jason Yosinsky, et al., 'The surprising creativity of digital evolution:
A collection of anecdotes from the Evolutionary Computation and
Artificial Life Research communities', ArXiv, March 2018 https://arxiv.org/
pdf/1803.03453v1.pdf
2. Christian Szegedy, et al., 'Explaining and harnessing adversarial examples'
https://arxiv.org/pdf/1412.6572v3.pdf?loc=contentwell&lnk=a-2015-
paper&dom=section-9

18: What is rationality?

1. Eliezer Yudkowsky, 'What do I mean by rationality?', LessWrong
sequences, 16 March 2009 hhttps://www.readthesequences.com/What-Do-I-
Mean-By-Rationality
2. Ibid.
3. Eliezer Yudkowsky, 'Newcomb's problem and regret of rationality',
LessWrong sequences, 31 January 2008 https://www.lesswrong.com/
posts/6ddcsdA2c2XpNpE5x/newcomb-s-problem-and-regret-of-rationality
4. Musashi Miyamoto, *The Book of Five Rings*, c.1645
5. Eliezer Yudkowsky, 'Why Truth? And . . ', LessWrong sequences, 27

November 2006 https://www.readthesequences.com/Why-Truth-And

6. Robert Nozick, 'Newcomb's problem and two principles of choice', *Essays in Honor of Carl G. Hempel* (Springer Netherlands, 1969) http://faculty.arts. ubc.ca/rjohns/nozick_newcomb.pdf

7. Eliezer Yudkowsky, 'Timeless Decision Theory', 2010 https://intelligence. org/files/TDT.pdf

19: Bayes' theorem and optimisation

1. https://en.wikipedia.org/wiki/Thomas_Bayes

2. Thomas Bayes, *An Essay towards solving a Problem in the Doctrine of Chances*, 1763

3. 'An intuitive explanation of Bayes' theorem', LessWrong sequences, 1 January 2003 https://www.readthesequences.com/An-Intuitive-Explanation-Of-Bayess-Theorem

4. Ward Casscells, Arno Schoenberger and Thomas Graboys, 'Interpretation by physicians of clinical laboratory results', *New England Journal of Medicine*, vol. 299, 1978, pp. 999–1001

5. 'Searching for Bayes-Structure', LessWrong sequences, 28 February 2008 https://www.readthesequences.com/SearchingForBayesStructure

6. Fred Hoyle, 'Hoyle on Evolution', *Nature*, vol. 294, No 5837 (12 November 1981), p. 105

7. Eliezer Yudkowsky, 'How much evidence does it take?', LessWrong sequences, 2007 https://www.readthesequences.com/How-Much-Evidence-Does-It-Take

8. Ibid.

20: Utilitarianism: shut up and multiply

1. Mason Hartman, @webdevmason, 2 April 2018 https://twitter.com/ webdevMason/status/980861298387836928

2. 'Extracts from Bentham's Commonplace Book', in *10 Works of Jeremy Bentham* (John Bowring, 1843), p. 141

3. Eliezer Yudkowsky, 'Torture vs dust specks', LessWrong sequences, 2008 https://www.lesswrong.com/posts/3wYTFWY3LKQCnAptN/torture-vs-dust-specks

4. Ibid.

5. Eliezer Yudkowsky, 'Circular altruism', LessWrong sequences, 2008 https://www.lesswrong.com/posts/4ZzefKQwAtMo5yp99/circular-altruism#uWXxEmfea9WFJmMSk

6. For instance, Alastair Norcross of the University of Colorado in his paper 'Comparing harms: Headaches and human lives', 1997 http://spot.colorado. edu/~norcross/Comparingharms.pdf

7. Derek Parfit, *Reasons and Persons* (OUP, 1984), p. 388

8. Eliezer Yudkowsky, 'The lifespan dilemma', LessWrong sequences, 2009

https://www.lesswrong.com/posts/9RCoE7jmmvGd5Zsh2/the-lifespan-dilemma

9. Eliezer Yudkowsky, 'Ends don't justify means (among humans)', LessWrong sequences, 2009 https://www.readthesequences.com/EndsDontJustifyMeansAmongHumans

10. Eliezer Yudkowsky, 'One life against the world', LessWrong sequences, 2007 https://www.lesswrong.com/posts/xiHy3kFni8nsxfdcP/one-life-against-the-world

21: What is a 'bias'?

1. Rob Bensinger, 'Biases: An introduction', LessWrong sequences, 2015 https://www.readthesequences.com/Biases-An-Introduction

2. Ibid.

22: The availability heuristic

1. David Anderson QC, 'The Terrorism Acts in 2011: Report of the Independent Reviewer on the Operation of the Terrorism Act 2000 and Part 1 of the Terrorism Act 2006', 2012 https://terrorismlegislationreviewer.independent.gov.uk/wp-content/uploads/2013/04/report-terrorism-acts-2011.pdf

2. Sarah Lichtenstein, et al., 'Judged frequency of lethal events,' *Journal of Experimental Psychology: Human Learning and Memory*, vol. 4(6), 1978, pp. 551–78 doi:10.1037/0278-7393.4.6.551

3. Elezier Yudkowsky, 'Availability', LessWrong sequences, 2008 https://www.readthesequences.com/Availability

4. Garrick Blalock, et al., 'Driving fatalities after 9/11: A hidden cost of terrorism', *Applied Economics*, vol. 41, issue 14, 2009 http://blalock.dyson.cornell.edu/wp/fatalities_120505.pdf

23: The conjunction fallacy

1. Eliezer Yudkowsky, 'Burdensome details', LessWrong sequences, 2007 https://www.readthesequences.com/Burdensome-Details

2. Amos Tversky and Daniel Kahneman, 'Judgments of and by Representativeness', in *Judgment Under Uncertainty: Heuristics and Biases*, ed. Daniel Kahneman, Paul Slovic and Amos Tversky (CUP, 1982), pp. 84–98

3. A. Tversky and D. Kahneman, 'Extensional versus intuitive reasoning: The conjunction fallacy in probability judgment', *Psychological Review*, vol. 90, 1983, pp. 293–315

24: The planning fallacy

1. Eliezer Yudkowsky, 'Planning fallacy', LessWrong sequences, 2007 https://www.readthesequences.com/Planning-Fallacy

2. Roger Buehler, Dale Griffin and Michael Ross, 'It's about time: Optimistic predictions in work and love,' *European Review of Social Psychology*, vol. 6(1), 1995, pp. 1–32 doi:10.1080/14792779343000112

3. Ian R. Newby-Clark, et al., 'People focus on optimistic scenarios and disregard pessimistic scenarios while predicting task completion times,' *Journal of Experimental Psychology: Applied*, vol. 6(3), 2000, pp. 171–82 doi:10.1037/1076-898X.6.3.171

4. Roger Buehler, Dale Griffin and Michael Ross, 'Exploring the "planning fallacy": Why people underestimate their task completion times,' *Journal of Personality and Social Psychology*, vol. 67(3), 1994, pp. 366–81 doi:10.1037/0022-3514.67.3.366

5. Roger Buehler, Dale Griffin and Michael Ross, 'Inside the planning fallacy: The causes and consequences of optimistic time predictions,' in Thomas Gilovich, Dale Griffin and Daniel Kahneman (eds), *Heuristics and Biases: The Psychology of Intuitive Judgment* (CUP, 2012), pp. 250–70

6. Yudkowsky, 'Planning fallacy' https://www.readthesequences.com/Planning-Fallacy

25: Scope insensitivity

1. Elezier Yudkowsky, 'Scope insensitivity', LessWrong sequences, 2008 https://www.readthesequences.com/ScopeInsensitivity

2. William H. Desvousges, et al., *Measuring Nonuse Damages Using Contingent Valuation: An Experimental Evaluation of Accuracy* (RTI Press, 1992) https://www.rti.org/sites/default/files/resources/bk-0001-1009_web.pdf

3. Richard T. Carson and Robert Cameron Mitchell, 'Sequencing and nesting in contingent valuation surveys', *Journal of Environmental Economics and Management*, vol. 28(2), 1995, pp. 155–73 doi:10.1006/jeem.1995.1011

4. Daniel Kahneman, Ilana Ritov and Daniel Schkade, 'Economic preferences or attitude expressions?: An analysis of dollar responses to public issues', *Journal of Risk and Uncertainty*, vol.19, issue 1–3, 1999, pp. 203–35 doi:10.1007/978-94-017-1406-8_8

5. David Fetherstonhaugh, et al., 'Insensitivity to the value of human life: A study of psychophysical numbing', *Journal of Risk and Uncertainty*, vol. 14(3), 1997, pp. 283–300 doi:10.1023/A:1007744326393

6. Rebecca Smith, '"Revolutionary" breast cancer drug denied on NHS over cost: NICE', *Daily Telegraph*, 8 August 2014

26: Motivated scepticism, motivated stopping and motivated continuation

1. Jonathan Haidt, *The Righteous Mind: Why Good People Are Divided by Politics and Religion* (Penguin, 2012), p. 98

2. Eliezer Yudkowsky, 'Motivated stopping and motivated continuation', LessWrong sequences, 2007 https://www.lesswrong.com/posts/

L32LHWzy9FzSDazEg/motivated-stopping-and-motivated-continuation

3. R.A. Fisher, 'Lung cancer and cigarettes', *Nature*, vol. 182, 12 July 1958, p. 108 https://www.york.ac.uk/depts/maths/histstat/fisher275.pdf

4. F. Yates and K. Mather, 'Ronald Aylmer Fisher 1890–1962', *Biographical Memoirs of Fellows of the Royal Society*, vol. 9, 1963, pp. 91–129 doi:10.1098/rsbm.1963.0006.

27: A few others, and the most important one

1. Eliezer Yudkowsky, 'Illusion of transparency: Why no one understands you', LessWrong sequences, 2007 https://www.readthesequences.com/Illusion-Of-Transparency-Why-No-One-Understands-You

2. Boaz Keysar, 'The illusory transparency of intention: Linguistic perspective taking in text', *Cognitive Psychology*, vol. 26(2), 1994, pp. 165–208 doi:10.1006/cogp.1994.1006

3. Eliezer Yudkowsky, 'Hindsight devalues science', LessWrong sequences, 2007 https://www.readthesequences.com/Hindsight-Devalues-Science

4. 'Did you know it all along?', excerpt from David G. Meyers, *Exploring Social Psychology* (McGraw-Hill, 1994), pp. 15–19 https://web.archive.org/web/20180118185747/https://musiccog.ohio-state.edu/Music829C/hindsight.bias.html

5. Eliezer Yudkowsky, 'The affect heuristic', LessWrong sequences, 2008 https://www.readthesequences.com/TheAffectHeuristic

6. Eliezer Yudkowsky, 'The halo effect', LessWrong sequences, 2008 https://www.readthesequences.com/TheHaloEffect

7. Eliezer Yudkowsky, 'Knowing about biases can hurt people', LessWrong sequences, 2008 https://www.readthesequences.com/Knowing-About-Biases-Can-Hurt-People

8. Ibid.

28: Thinking probabilistically

1. Tetlock quoted in Dan Gardner, *Future Babble* (Virgin Books, 2011), p. 24

2. Ibid., p. 25

3. Isaiah Berlin, *The Hedgehog and the Fox: An Essay on Tolstoy's View of History* (Princeton Press, 1953)

29: Making beliefs pay rent

1. Eliezer Yudkowsky, 'Disputing definitions', LessWrong sequences, 2008 https://www.readthesequences.com/Disputing-Definitions

2. Eliezer Yudkowsky, 'Making beliefs pay rent (in anticipated experiences)', LessWrong sequences, 2008 https://www.readthesequences.com/Making-Beliefs-Pay-Rent-In-Anticipated-Experiences

30: Noticing confusion

1. Eliezer Yudkowsky, 'Fake explanations', LessWrong sequences, 2008
https://www.readthesequences.com/Fake-Explanations

31: The importance of saying 'Oops'

1. Eliezer Yudkowsky, 'The importance of saying "Oops"',
LessWrong sequences, 2008 https://www.readthesequences.com/
TheImportanceOfSayingOops
2. Ibid.

32: The semi-death of LessWrong

1. Riciessa, 'LessWrong analytics, February 2009 to January 2017' https://
www.lesswrong.com/posts/SWNn53RryQgTzT7NQ/lesswrong-analytics-
february-2009-to-january-2017
2. Scott Alexander, 'A History of the Rationalist community', Reddit, 2017
https://www.reddit.com/r/slatestarcodex/comments/6tt3gy/a_history_of_
the_rationality_community/
3. Scott Alexander, 'Mapmaker, mapmaker, make me a map', 2014 https://
slatestarcodex.com/2014/09/05/mapmaker-mapmaker-make-me-a-map/

33: The IRL community

1. Zvi Mowshowitz, 'The thing and the symbolic representation of the thing',
2015 https://thezvi.wordpress.com/2015/06/30/the-thing-and-the-symbolic-
representation-of-the-thing/
2. Sara Constantin, 'Lessons learned from MetaMed', 2015 https://docs.
google.com/document/d/1HzZd3jsG9YMU4DqHc62mMqKWtRer_
KqFpiaeN-QirlI/edit
3. 2016 LessWrong diaspora survey results http://www.jdpressman.
com/public/lwsurvey2016/Survey_554193_LessWrong_Diaspora_2016_
Survey%282%29.pdf
4. SSC survey results 2018 http://slatestarcodex.com/2018/01/03/ssc-survey-
results-2018/

34: Are they a cult?

1. Unknown author, 'Our phyg is not exclusive enough', LessWrong, 2012
https://www.lesswrong.com/posts/hxGEKxaHZEKT4fpms/our-phyg-is-not-
exclusive-enough
2. reddragdiva.tumblr.com, https://reddragdiva.tumblr.com/
post/172165021858/some-charities-are-more-effective-than-others-and
3. Eliezer Yudkowsky, 'Every cause wants to be a cult', LessWrong, 2007
https://www.lesswrong.com/posts/yEjaj7PWacno5EvWa/every-cause-wants-

to-be-a-cult

4. 'Transcription of Eliezer's January 2010 video Q&A', LessWrong, 2010
https://www.lesswrong.com/posts/YduZEfz8usGbJXN4x/transcription-of-eliezer-s-january-2010-video-q-and-a

5. MIRI Independent Auditors Report for 2016, https://intelligence.org/wp-content/uploads/2012/06/Independent-Auditors-Report-for-2016.pdf

6. Scott Alexander, 'The Noncentral Fallacy: The worst argument
in the world', LessWrong, 2012 https://www.lesswrong.com/posts/yCWPkLi8wJvewPbEp/the-noncentral-fallacy-the-worst-argument-in-the-world

7. SSC 2018 survey results http://slatestarcodex.com/2018/01/03/ssc-survey-results-2018/

8. LessWrong 2014 survey results http://lesswrong.com/lw/lhg/2014_survey_results/

9. Elizabeth Sheff, 'How many polyamorists are there in the US?' *Psychology Today*, 9 May 2014 https://www.psychologytoday.com/us/blog/the-polyamorists-next-door/201405/how-many-polyamorists-are-there-in-the-us

10. M.L. Haupert et al., 'Prevalence of experiences with
consensual nonmonogamous relationships: Findings from two
national samples of single Americans', *Journal of Sex & Marital Therapy*, vol. 43, issue 5, 2017 https://www.tandfonline.com/doi/abs/10.1080/0092623X.2016.1178675?journalCode=usmt20

11. Brendan Shucart, 'Polyamory by the numbers', *Advocate*, 1 August 2016
https://www.advocate.com/current-issue/2016/1/08/polyamory-numbers

35: You can't psychoanalyse your way to the truth

1. Scott Alexander, 'Is everything a religion?', 2015 http://slatestarcodex.com/2015/03/25/is-everything-a-religion/

2. John Horgan, 'The consciousness conundrum', *IEEE Spectrum*, 1 June
2008 https://spectrum.ieee.org/biomedical/imaging/the-consciousness-conundrum

3. John Horgan, 'AI visionary Eliezer Yudkowsky on the Singularity,
Bayesian brains and closet goblins', *Scientific American*, 1 March 2016 https://blogs.scientificamerican.com/cross-check/ai-visionary-eliezer-yudkowsky-on-the-singularity-bayesian-brains-and-closet-goblins/

36: Feminism

1. Scott Aaronson, comment #171 under blog post 'Walter
Lewin', Shtetl-Optimised, 2015 https://www.scottaaronson.com/blog/?p=2091#comment-326664

2. Amanda Marcotte, 'MIT professor explains: The real oppression is
having to learn to talk to women', RawStory, 2014 https://www.rawstory.com/2014/12/mit-professor-explains-the-real-oppression-is-having to-learn-

to-talk-to-women/

3. Scott Alexander, 'Untitled', 2015 http://slatestarcodex.com/2015/01/01/untitled/

4. Russell Clark and Elaine Hatfield, 'Gender differences in receptivity to sexual offers', *Journal of Psychology & Human Sexuality*, vol. 2(1), 1989 https://www.tandfonline.com/doi/abs/10.1300/J056v02n01_04

5. Vashte Galpin, 'Women in computing around the world: An initial comparison of international statistics', *ACM SIGCSE Bulletin*, vol. 34(2), June 2002, pp. 94–100 http://homepages.inf.ed.ac.uk/vgalpin1/ps/Gal02a.pdf

6. Elizabeth Weise, 'Tech: Where the women and minorities aren't', *USA Today*, 15 August 2014 https://eu.usatoday.com/story/tech/2014/05/29/silicon-valley-tech-diversity-hiring-women-minorities/9735713/

7. NCWIT, 'Girls in IT: The facts infographic', National Center for Women and IT, 30 November 2012 https://www.ncwit.org/infographic/3435

8. Beth Gardiner, 'Computer coding: It's not just for boys', *New York Times*, 7 March 2013 https://www.nytimes.com/2013/03/08/technology/computer-coding-its-not-just-for-boys.html

9. Kathy A. Krendl, Mary C. Broihier and Cynthia Fleetwood, 'Children and computers: Do sexrelated differences persist?', *Journal of Communication*, vol. 39(3), 1 September 1989, pp. 85–93 https://doi.org/10.1111/j.1460-2466.1989.tb01042.x

10. Lily Shashaani, 'Gender differences in computer attitudes and use among college students', *Journal of Educational Computing Research*, vol. 16, issue 1, 1 January 1997 http://journals.sagepub.com/doi/abs/10.2190/Y8U7-AMMA-WQUT-R512?journalCode=jeca

11. Richard A. Lippa, 'Gender differences in personality and interests: When, where, and why?', *Social and Personality Psychology Compass*, vol. 4, issue 11, 20 October 2010 https://doi.org/10.1111/j.1751-9004.2010.00320.x

12. Scott Alexander, 'Contra Grant on exaggerated differences', 2017 http://slatestarcodex.com/2017/08/07/contra-grant-on-exaggerated-differences/

13. 'The state of medical education and practice in the UK', General Medical Council, 2016 https://www.gmc-uk.org/-/media/documents/SOMEP_2016_Full_Report_Lo_Res.pdf_68139324.pdf

14. 'Higher Education Student Statistics: UK, 2016/17 – Subjects studied', HESA, 2018 https://www.hesa.ac.uk/news/11-01-2018/sfr247-higher-education-student-statistics/subjects

15. 'Association of American Medical Colleges 2015 Report on Residents', AAMC, 2015 https://www.aamc.org/data/484710/report-on-residents.html

16. 'The state of medical education and practice in the UK', General Medical Council, 2017 https://www.gmc-uk.org/-/media/about/somep-2017-final-full.pdf?la=en&hash=3FC4B6C2B7EBD840017B908DBF0328CD840640A1

17. Scott Alexander, 'Untitled', 2015 http://slatestarcodex.com/2015/01/01/untitled/

18. 'Full leaked Googlers' conversations regarding the Google memo', reddit.

com/r/KotakuInAction http://archive.is/wUBb5#selection-2283.0-2301.9
19. Emily Gorcenski, 'Will this make people afraid to share their thoughts?
Yes. Shitty people should be afraid to share their fascist thoughts.' Twitter
https://twitter.com/EmilyGorcenski/status/893973537941327876
20. James Damore, 'The document that got me fired from Google', 2017
https://firedfortruth.com/
21. Paul Lewis, '"I see things differently": James Damore on his autism and
the Google memo', *Guardian*, 17 November 2017 https://www.theguardian.
com/technology/2017/nov/16/james-damore-google-memo-interview-
autism-regrets

37: The Neoreactionaries

1. 2016 LessWrong diaspora survey results http://www.jdpressman.
com/public/lwsurvey2016/Survey_554193_LessWrong_Diaspora_2016_
Survey%282%29.pdf
2. LessWrong diaspora survey 2016 http://www.jdpressman.com/public/
lwsurvey2016/analysis/general_report.html
3. Scott Alexander, 'You're probably wondering why I've called you here
today', 2013 http://slatestarcodex.com/2013/02/12/youre-probably-wondering-
why-ive-called-you-here-today/
4. Scott Alexander, 'SSC Endorses Clinton, Johnson, or Stein', 2016 http://
slatestarcodex.com/2016/09/28/ssc-endorses-clinton-johnson-or-stein/
5. SSC survey results 2018 http://slatestarcodex.com/2018/01/03/ssc-survey-
results-2018/

38: The Effective Altruists

1. Peter Singer, 'Famine, affluence, and morality', *Philosophy and Public
Affairs*, vol. 1(1), Spring 1972, pp. 229–43 (rev. edn) https://www.utilitarian.
net/singer/by/1972----.htm
2. Benjamin Todd, 'Earning to give', 80,000 Hours, 2017 https://80000hours.
org/articles/earning-to-give/
3. 'What we can achieve', Giving What We Can https://www.
givingwhatwecan.org/get-involved/what-we-can-achieve/
4. Giving USA 2018: The Annual Report on Philanthropy for the Year 2017
https://givingusa.org/giving-usa-2018-americans-gave-410-02-billion-to-
charity-in-2017-crossing-the-400-billion-mark-for-the-first-time/
5. The 2014 survey of Effective Altruists, Centre for Effective Altruism http://
effective-altruism.com/ea/gb/the_2014_survey_of_effective_altruists_results/
6. EA 2017 survey, Centre for Effective Altruism https://rtcharity.org/tag/
ea-survey-2017/
7. LessWrong diaspora survey 2016 http://www.jdpressman.com/public/
lwsurvey2016/Survey_554193_LessWrong_Diaspora_2016_Survey282%29.pdf
8. Scott Alexander, 'Nobody is perfect, everything is commensurable', 2014

http://slatestarcodex.com/2014/12/19/nobody-is-perfect-everything-is-commensurable/

9. 'About chickens', Compassion in World Farming, 2017 https://www.ciwf.org.uk/farm-animals/chickens/

10. Scott Alexander, 'Vegetarianism for meat-eaters', 2015 http://slatestarcodex.com/2015/09/23/vegetarianism-for-meat-eaters/

11. Wild-Animal Suffering Research, https://was-research.org/mission/

12. 'Is there suffering in fundamental physics?', Foundational Research Institute http://reducing-suffering.org/is-there-suffering-in-fundamental-physics/

13. 'Impossible Foods – R&D Investment', OpenPhil 2016 https://www.openphilanthropy.org/focus/us-policy/farm-animal-welfare/impossible-foods

14. Katie Strick, 'This is what the "bleeding" vegan burger at Mildreds is really like', *London Evening Standard*, 21 February 2018 https://www.standard.co.uk/go/london/restaurants/this-is-what-the-bleeding-vegan-burger-at-mildreds-is-really-like-a3772061.html

39: EA and AI

1. Benjamin Todd, 'Why, despite global progress, humanity is probably facing its most dangerous time ever', 80,000 Hours https://80000hours.org/articles/extinction-risk/

2. Holden Karnofsky, 'Potential risks from advanced artificial intelligence: The philanthropic opportunity', OpenPhil https://www.openphilanthropy.org/blog/potential-risks-advanced-artificial-intelligence-philanthropic-opportunity#Tractability

3. 'Machine Intelligence Research Institute – general support', OpenPhil 2016 https://www.openphilanthropy.org/focus/global-catastrophic-risks/potential-risks-advanced-artificial-intelligence/machine-intelligence-research-institute-general-support

4. 'Our progress in 2017 and plans for 2018', OpenPhil https://www.openphilanthropy.org/blog/our-progress-2017-and-plans-2018

5. reddragdiva.tumblr.com, https://reddragdiva.tumblr.com/post/172165021858/some-charities-are-more-effective-than-others-and

6. Dylan Matthews, 'I spent a weekend at Google talking with nerds about charity. I came away . . . worried', Vox, 10 August 2015 https://www.vox.com/2015/8/10/9124145/effective-altruism-global-ai

7. Ben Kuhn, 'Some stories about comparative advantage', December 2014 https://www.benkuhn.net/advantage

8. Edward Miguel and Michael Kremer, 'Worms: Identifying impacts on education and health in the presence of treatment externalities', *Econometrica*, vol. 72(1), January 2004, pp. 159–217

9. D.C. Taylor-Robinson, N. Maayan, K. Soares-Weiser, S. Donegan and P.

Garner, 'Deworming drugs for soil-transmitted intestinal worms in children: Effects on nutritional indicators, haemoglobin, and school performance', Cochrane Database of Systematic Reviews, 23 July 2015 (CD000371)
10. Philip Oltermann, 'Greenpeace loses £3m in currency speculation', Guardian, 16 June 2014 https://www.theguardian.com/environment/2014/jun/16/greenpeace-loses-3m-pounds-currency-speculation
11. 'Introducing OpenAI', OpenAI.com 2015 https://blog.openai.com/introducing-openai/

40: What are they doing to stop the AI apocalypse?

1. D. Amodei, C. Olah, J. Steinhardt, P. Christiano, J. Schulman and D. Mane, 'Concrete problems in AI safety', technical report, 25 July 2016 arXiv:1606.06565v2 (cs.AI)
2. Paul Christiano, et al., 'Deep reinforcement learning from human preferences', OpenAI https://blog.openai.com/deep-reinforcement-learning-from-human-preferences/
3. Paul Christiano, 'Capability amplification', Medium https://ai-alignment.com/policy-amplification-6a70cbee4f34

41: The internal double crux

1. Duncan Sabien, 'Double crux – A strategy for resolving disagreement', LessWrong, 2017 https://www.lesswrong.com/posts/exa5kmvopeRyfJgCy/double-crux-a-strategy-for-resolving-disagreement

42: Life, the universe and everything

1. Bostrom and Müller, 'Future progress in artificial intelligence' https://nickbostrom.com/papers/survey.pdf
2. Bostrom, Superintelligence, p. 115
3. 'Preparing for the Future of Artificial Intelligence', Executive Office of the President, National Science and Technology Council Committee on Technology, October 2016 https://obamawhitehouse.archives.gov/sites/default/files/whitehouse_files/microsites/ostp/NSTC/preparing_for_the_future_of_ai.pdf